Praise for *Math-Positive Mindsets . . .*

Although I am a business leader who loves to learn, I have always struggled with math and frequently consult my finance teams to assist in simplifying complex financial concepts and math models. This book gives me courage in the area of math, helps me know that I am not alone, and provides incredible tools for me to use with my kids to avoid many of the mistakes and anxieties I developed learning math. I cannot thank the author, Carrie Cutler, enough. Well done!

—Bryant P. Castleton, Vice President of Stanson Business Development, Georgia

The expectation that we help children develop math-positive mindsets requires that we, their parents and teachers, cultivate such mindsets for ourselves as well. The author begins by issuing the challenge, and she doesn't stop until she gives us practical answers to numerous questions we are likely to have plus concrete examples to help us internalize the "how" as well as the "why." This book is a must-read for those who impact young children's lives mathematically.

—Eula Ewing Monroe, Professor Emeritus, Brigham Young University, Utah

I love math and, as a homeschooling mother of six, it breaks my heart when my children struggle with the language and procedures associated with math. After reading the first chapter of *Math-Positive Mindsets*, I turned to my seven-year-old and used just *one* of the amazing tips Dr. Cutler offers. Instead of crying, we both were grinning by the end of the lesson—about fractions! Thank you so much for creating this! Look out, math! We're coming for you!

—Bonnie R. Paulson, *USA Today* bestselling author

Math-Positive Mindsets is an excellent resource to help parents, teachers, and students reframe the way they think about math. The book instills a positive approach to math, opening minds to new techniques for teaching and understanding math concepts. It is highly beneficial to our program's teachers as they introduce math to our students in ways that are unique and developmentally appropriate.

—Leslie Shaw, Preschool Director, Red Bridge United Methodist Church, Missouri

Math has always come easily to me, so I'm surprised that my children have a difficult time understanding and enjoying the subject. *Math-Positive Mindsets* gives me a greater understanding of my children's struggle as well as offers ideas for how to assist them in developing a more positive math attitude. This book is an invaluable resource for parents wanting to help their child master math concepts in a positive way.

—Heather Anderson, mom of two, Colorado

Featuring authentic questions and concerns in a friendly Q&A format, *Math-Positive Mindsets* is an essential resource for anyone who has experienced math anxiety. Carrie Cutler writes understandable, clear, and time-tested answers with appropriate activities. As a teacher of teachers, a parent, and a longtime math teacher, I strongly recommend this resource for anyone who works with children.

—Juanita Copley, author of *The Young Child and Mathematics*

Math-Positive Mindsets is essential for parents who feel that they cannot help their children with math homework and teachers who want to help parents help students. Change your mind from "I don't know what's going on" to "I can learn what's going on" and be the difference in children's math education!

—Jesus Sanchez, bilingual math teacher, Texas

As parents who are semiterrified of having to help our children with math homework, we read and then reread the chapters in *Math-Positive Mindsets*. This book helps us understand what is happening in our children's classroom from the teacher's perspective. Our favorite parts are the real-life examples Carrie shares, helping us understand math and kids better. This is required reading for parents who want to help their children be successful and have a positive math experience.

—Paul and Tiffany Ivanovsky, MyLitter.com, featured on the *Today* show, *Rachael Ray* show, and many more national outlets

Cutler provides a beautiful framework that supports and fosters a math-positive classroom for new and experienced educators alike. By offering tools, tips, and implementation strategies, *Math-Positive Mindsets* is a resource that will empower all teachers . . . and their students.

—Graham Fletcher, math specialist, gfletchy.com, Twitter: @gfletchy

As a parent of a child attending Title I math, I found *Math-Positive Mindsets* to be a great resource for finding positive ways to make math accessible to a child who learns a little differently.

—Annie Javadi, mom of five, Oregon

Cutler's writing style makes math accessible to parents with easy-to-understand explanations, fun activities, and simple discussion starters. With *Math-Positive Mindsets*, any parent can be a math-positive parent.

—Jane Clayson Johnson, author of *Silent Souls Weeping: Sharing Stories, Finding Hope*

Math-Positive Mindsets has really opened my eyes to teaching math in natural, everyday ways. As a nanny to young children, I'm so grateful to have finally found a book that takes the complexity out of teaching math.

—Amy Scoville, nanny, South Carolina

Math-Positive Mindsets is an overdue treasure, invaluable for anyone who teaches children math—from the classroom teacher to the parent sitting at a kitchen table. The book's format is helpful, not intimidating, through friendly "Pause"ative Boxes, Teaching Tips, and Q&A examples. As a parent, former school principal, and teacher, I am excited to use this book for professional learning in our district and beyond. Thank you, Carrie, for inspiring all to love math the way you do!

—Alicia Rudd, District Curriculum Coordinator, Nebo School District, Utah

MATH-POSITIVE MINDSETS

Growing a Child's Mind without Losing Yours

GRADES K–5

Carrie S. Cutler

Math Solutions
Boston, Massachusetts, USA

Math Solutions
www.mathsolutions.com

Library of Congress Control Number: 2019955416

ISBN-13: 978-1-935099-84-0
ISBN-10: 1-935099-84-1

Math Solutions is a division of Houghton Mifflin Harcourt.

Executive Editor: Jamie Ann Cross
Production Manager: Denise A. Botelho
Editorial Assistant: Kirby Sandmeyer
Cover design: Vicki Tagliatela, DandiLion Designs
Cover photos: (front) BraunS, LightFieldStudios, and artJazz; (back) Cart icon made by Freepik from www.flaticon.com
Author photo: KWW Photography
Interior design and composition: Publishers' Design and Production Services, Inc.

Printed in the United States of America.
1 2 3 4 5 6 7 8 9 10 0304 26 25 24 23 22 21 20
4510006791 ABCDE

A Message from Math Solutions

We at Math Solutions believe that teaching math well calls for increasing our understanding of the math we teach, seeking deeper insights into how students learn mathematics, and refining our lessons to best promote students' learning.

Math Solutions shares classroom-tested lessons and teaching expertise from our faculty of professional learning consultants as well as from other respected math educators. Our publications are part of the nationwide effort we've made since 1984 that now includes:

- more than five hundred face-to-face professional learning programs each year for teachers and administrators in districts across the country;
- professional learning books that span all math topics taught in kindergarten through high school;
- videos for teachers and for parents that show math lessons taught in actual classrooms;
- on-site visits to schools to help refine teaching strategies and assess student learning; and
- free online support, including grade-level lessons, book reviews, inservice information, and district feedback, all in our Math Solutions Online Newsletter.

For information about all of the products and services we have available, please visit our website at *www.mathsolutions.com.* You can also contact us to discuss math professional development needs by calling (877) 234-7323 or by sending an email to *info@mathsolutions.com.*

We're always eager for your feedback and interested in learning about your particular needs. We look forward to hearing from you.

To the Cutler children—

William Royal, Sybil Adrienne, Charles Shumway, James Zebulon, Quinn Helaman Hollands, McGregor Fitzgerald, Knox Dalley, and especially to Duncan Stewart, my out-of-the-rectangular prism thinker.

May you always see your infinite potential— mathematical and otherwise.

Brief Contents

How to Access Online Resources
To access the downloadable Quick Reference Chart for each chapter, please visit mathsolutions.com/myonlineresources and register for an account (even if you have one with our bookstore). Use key code **MPM** to register this product and download the charts.

Contents

(continued)

(continued)

(continued)

(continued)

(continued)

Acknowledgments

This book is a labor of love. Love for mathematics. Love for children. And love for the parents and teachers who strive to join math and children together in math-positive ways. It took an immense amount of effort to complete and would not exist without the invaluable contributions of many cheerleaders and champions.

I must start by thanking my husband, Chris Cutler, who supported me unfailingly throughout the writing of this book. Thank you, my dearest darling Christopher, for reading drafts, making suggestions, offering encouragement, and giving me many, many Saturdays of kid-free time to write.

Sincere thanks to the talented editorial, production, and marketing team at Math Solutions. Jamie Cross, Kirby Sandmeyer, Denise Botelho, Kelli Cook, Diane Reynolds, Lisa Bush, Patricio Dujan, Patty Clark, Amy Mayfield, Renee Hammonds, and of course Marilyn Burns. All the Math Solutions friends I've made over the past several years have contributed expertise and encouragement throughout the process of writing, revisiting, revising, and refining this book. Thank you for giving me a platform for my silly stories and impassioned pleas to improve mathematics education for children and the adults who love them.

Two professional mentors gave me a vision for myself beyond what I may have otherwise imagined. Dr. Eula Ewing Monroe of Brigham Young University co-authored my first published article and has included me as a chapter author and co-presenter at National Council of Teachers of Mathematics convenings for more than fifteen years. Dr. Juanita Copley of the University of Houston steered me toward early childhood education as a niche, set me up as an education consultant, and let me treasure hunt in her garage when she moved. Thank you for showing me how I could be more of me.

My parents, Boyd and Tina Stewart, two of the best writers I know, who taught me their craft as we sat across the kitchen table on the farm in Idaho. My in-laws, Craig and Cheryl Cutler, for your interest in my career as well as the copious grandchildren I provided you. My siblings, Jennifer Griffith, Bill Stewart, and Adrienne Johnson for being supportive of and patient with my math obsession—even when I get overexcited about prime numbers.

Finally, thank you to my fellow parents and teachers who chuckled and nodded upon hearing my original title for this book, *How to Help Your Kids with Their Math Homework without Strangling Them*. You affirmed this book was so very needed. May this book help you continue to grow in your math-positive mindset!

How to Use This Resource

WHAT IS A MATH-POSITIVE MINDSET?

As the mom of a large family, I spend a lot of time pushing grocery carts. When I ran into a friend at the store recently, she asked how my semester teaching Elementary Math Methods was going. I beamed while telling her how much I love teaching teachers how to teach math. She listened politely but bristled, "I was never any good at math, and when my kids need help with their math homework, I just about pull my hair out." She continued for several minutes detailing the horrors of sitting across the kitchen table from a struggling child and feeling like a struggling parent.

I offered a few words of encouragement, but the conversation left me wondering: Could other parents feel as frustrated and confused?

Curious, I asked my social media friends to share their experiences helping their children with math homework. I got dozens of responses—almost all negative—such as:

- Today I reviewed one of my son's math tests and literally felt my right eye twitching and my heart racing. Anxiety does not begin to describe it!
- I freaking hate math!
- I am not exaggerating when I say that my brain shuts off where math homework is involved. Right now, my twin girls are only in kindergarten, so I can handle helping them, but I am horrified to think that I will have to assist in stuff I don't understand. Guess I better save my money for a tutor.

Wow, I didn't realize that so many of my friends have what Dr. Carol Dweck would call a *fixed mindset* about math. A mindset, according to Dweck (2006), is a self-perception people hold about malleability and their brain's ability to grow and build intelligence. Putting the academic mumbo jumbo aside, a mindset is what you think about your thinking—if you believe you can improve your thinking and make it better. Simply put, a person may hold either a *growth* or a *fixed* mindset, believing they are "intelligent" or "unintelligent," "good at math" or "bad at math."

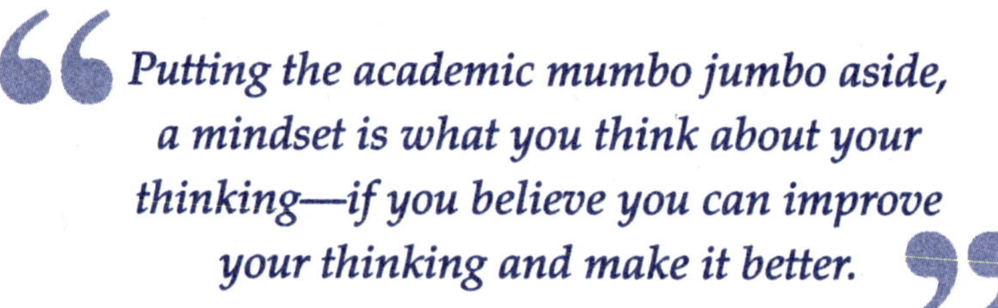

But why does this matter? Aren't we all just born good or bad at math? Nope. Research from Dweck and others (e.g., Boaler 2013; Yeager and Dweck 2012) shows that what we believe about our mathematics potential profoundly affects what we learn. If we have a *math-positive mindset*, we believe that hard work and effort, not natural talent alone, are what lead to mathematics achievement. We believe that struggling in math is a good thing, because struggling means we are trying to understand, not just trying to get by. We believe that mistakes in math create real space for learning and growing the brain.

If we have a math-positive mindset, we believe that hard work and effort, not natural talent alone, are what lead to mathematics achievement.

Research has disproven the long-held myth that only a select few individuals are "good at math." Everyone can do mathematics. Still, too many people fool themselves into thinking that they must be born with a mathematical mind. That only nerds can do math. That making mistakes means you're not good at math. That struggling to understand math means it's a lost cause.

Enough already!

WHO IS *MATH-POSITIVE MINDSETS* FOR?

Math is for everyone, and so is this resource! Too many adults (yes, even teachers) lack confidence in their own mathematical abilities. Math fear hinders their efforts to support and encourage children, even at the elementary-school level.

For those with children, I hope this book helps you evolve from a math-panicked parent to a *math-positive parent*. While it won't make you an instant math whiz, this book will set you on the path to a math-positive mindset. A math-positive parent keeps any lurking personal math anxiety in check. Math-positive parents speak positively about math. They help children work hard to understand math and support effort, not just grades. A math-positive parent creates an encouraging homework environment. Armed with persistence and patience, a math-positive parent provides help as appropriate. Math-positive parents may even feel safe saying that they learn *alongside* their children.

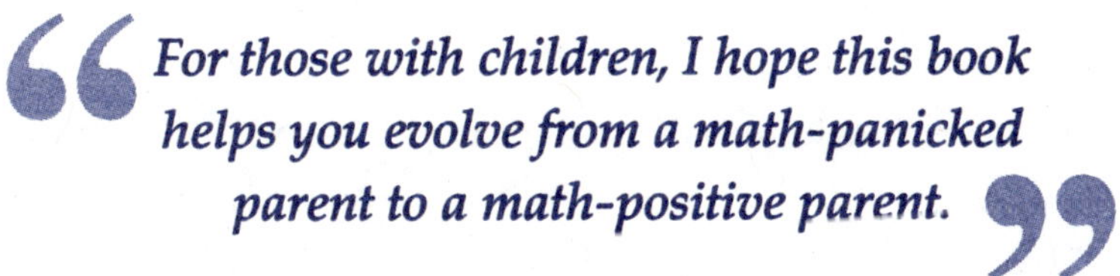

Teachers, this book offers activities, lesson ideas, and resources that will support your math-positive classroom culture. Carefully selected math tasks can produce the "just-right challenge" that pushes students beyond their comfort zones into the space where productive struggle generates deep learning. Too often, teachers with a fixed mindset about math mollify children by patting them on the head and saying,

"Don't worry, sweetie. I was never really good at math either." Or, "You're really more of a language arts person." Don't give children permission to dislike, fear, or avoid math. Rather, they will love, appreciate, and excel at math when they have the opportunity to feed off your math enthusiasm. The teaching tips sprinkled throughout this book help teachers replace fixed-mindset missives with math-positive prompts, pose challenging problems that push students to show persistence, and give practical advice on how to help students learn from their mistakes.

> ***Teachers, this book offers activities, lesson ideas, and resources that will support your math-positive classroom culture.***

WHY THIS RESOURCE?

I'm a college professor. I've hit the library and have written this book based on research from the National Council of Teachers of Mathematics and the National Association for the Education of Young Children. I've reviewed professional journals in child development and cognitive theory. I'm fluent in the best practices in education. I've taught teachers how to teach math for more than two decades. I present at national conferences and give workshops to teachers.

But most of all, I'm a mom.

When people ask me what I do, I don't even blink before spitting out one word, "laundry." Of course, that isn't all I do. But I wash, dry, fold, and put away at least thirteen loads a week. That's a lot of hours doing laundry. And, like you, I spend almost as many hours helping with homework.

Mercifully, I also carve out a few precious hours to study how kids learn math and how teachers can more effectively teach math. The two hobbies may seem completely disparate. But I have a vested interest in kids learning math and learning it well (hopefully while wearing clean underwear). I'm the mom of eight kids!

My fourteen-year-old son, Duncan, has a particularly nontraditional/unique/sometimes-befuddling approach to learning. Our hours across the kitchen table working together on homework can be precious, cherished moments, but also periodically make both of us pull our hair out. His struggles make me want to encourage all parents to hang in there. No matter how tense homework time can be, your child and you can build a math-positive mindset as you learn math together.

HOW THIS RESOURCE IS ORGANIZED

This book's question-and-answer (Q and A) format contains practical, plain-language responses that give instant access to helpful tips and explanations. Look to the Contents for a list of all questions addressed in the seven chapters in this book. In addition to the Q and A format, each chapter offers:

"Pause"ative Boxes

A play on the word *positive*, the information in "Pause"ative boxes builds on what is conveyed through the corresponding Q and A and tasks the reader with pausing and thinking more deeply about the topic. These sections can also be used to promote positive conversation in book study groups or professional learning communities for teachers and math coaches who work collaboratively to improve math instruction in schools.

"PAUSE"ATIVE BOX

Jo Boaler on Growth Mindset

Take a moment to pause and watch Jo Boaler's TEDx-Stanford video, "How You Can Be Good at Math, and Other Surprising Facts about Learning." In this video, Boaler, a professor of mathematics education at Stanford, shares what brain research about the growth mindset teaches us about learning and success in math. Wait 'til you hear what she says about mistakes in math! The findings of her research may surprise you.

An excerpt from a "Pause"ative Box in this book.

How to Access Online Resources

To access the downloadable Quick Reference Chart for each chapter, please visit mathsolutions.com/myonlineresources and register for an account (even if you have one with our bookstore). Use key code **MPM** to register this product and download the charts.

Teaching Tips

Teaching tips, sprinkled throughout, offer information specific to the teacher, from ideas for questions to ask to activities that support math-positive mindsets.

Wall of Mistakes

TEACHING TIP

Cultivate a growth mindset in your students' parents by including growth mindset activities at family math nights. Here's one idea: Have parents add to a Wall of Mistakes by writing their foibles on a sticky note and hanging it up to celebrate how errors help our brains grow.

An excerpt from a Teaching Tip section of this book.

Quick Reference Chart

Every chapter features a quick reference chart at the end; these charts summarize the key concerns in the chapter and are formatted so they can be printed or photocopied and displayed as an at-a-glance reminder as you contribute to building math-positive mindsets.

QUICK REFERENCE CHART

Common Concerns about Math Attitudes

This table is a quick reference for some of the *most common concerns* related to math attitudes. The table lists the concern, an explanation of the thinking behind it, and an idea that addresses the concern. It really is true that our attitude determines our effort, which in turn affects our outcomes. Work to improve your math-positive mindset by trying out the suggestions in the *What to Do* column.

Concern	Explanation	What to Do	Question Number(s) (to Learn More)
Math myths (you must be born good at math)	This is a lingering misconception. Brain research shows there is no such thing as a math gene.	Go online and find Carol Dweck's TEDx Talk entitled "The Power of Yet" and watch it with your children. Discuss how the brain of a person with a growth mindset responds to challenges and to mistakes.	1, 2

An excerpt from a Quick Reference Chart in this book.

Step I: Develop Your Math-Positive Mindset

To begin your evolution from math panicked to math positive, you should first explore what a math-positive mindset is and how to cultivate it. **Chapter 1** gleefully cheers, "All for math and math for all." Reading this chapter should convince you that math is for everyone and, with effort, every parent and teacher can develop the attitudes, tools, and practices that support children's learning.

Step II: Get to Know Your Math-Positive Team

The second step in building a math-positive mindset is to learn more about your math-positive team. For both the teacher and parent, this means getting to really know the math classroom—and getting to know each other! **Chapter 2** gives parents a glimpse into the secret world of teachers and their classrooms. OK, it's not quite as dramatic as that. But it's helpful for parents to understand why teachers do what they do. The chapter opens with suggestions on how to have successful parent-teacher conferences and how to word emails to teachers (and back to parents) without freaking them out. For parents, the chapter then moves into understanding why your child's math classroom might look and feel different from the classrooms you remember as a child. Among other things, this chapter explains why teachers emphasize problem solving (because we want children to *think* not just follow procedures) and why we ask children to explain their answers (because we're interested in their flexibility in math reasoning). For teachers, the chapter provides helpful tips for using sentence frames to guide students' reasoning and communication in math and alternatives to timed tests for assessing fact fluency. The chapter concludes with a couple of Q and A's on state standards.

Step III: Apply Your Math-Positive Mindset

Now that you're on your way to developing a math-positive mindset and you're working closely with your math-positive team (parents and teachers alike), it's time to support children's math learning by encouraging a positive math environment at home and taking a deeper dive into the math content being learned.

Chapter 3 explores ways to encourage math learning at home. It takes a look at homework and how to create a positive space for such, including what supplies to have on hand at home. It looks at what questions to ask to support children as they grapple with math and how to remain calm and resist the urge to take over (or give up!). For times when additional help might be needed, this chapter offers suggestions for adding a math tutor to your math-positive team and finding assistance online. The latter part of the chapter provides a list of children's books for encouraging math through reading at home or in school, as well as ideas for connecting math to everything done outside of school, such as grocery shopping.

Chapters 4 through 7 offer specific advice for the content areas in math—Number and Operations, Geometry, Measurement and Data, and Algebra. These chapters break down confusing concepts, tackle common misconceptions, and present ways parents and teachers can reinforce mathematical vocabulary while supporting conceptual learning over memorized procedures. They are filled with simple explanations for some of the tricky, modern incarnations of the math you learned in elementary school and plenty of math-positive, attitude-building activities.

HOW *MATH-POSITIVE MINDSETS* IS MEANT TO BE READ

I don't expect that you will read this book from cover to cover. Instead, I imagine that you'll pick it up during a particularly trying afternoon of a math lesson or fractions homework. I imagine that you'll refer to it for simple teaching ideas for reviewing basic addition facts. I imagine that you'll flip through it to find an anecdote that brings a smile to your face when you're at your wits' end. Some sections are better suited to parents and teachers of young children. Some are meant for older kids. Teachers can find tips throughout. Feel free to skip around. Look in the contents to find a question that matches your own. Or skim them all. I hope you'll find plenty of support in your efforts to help children.

> ***"Feel free to skip around. Look in the contents to find a question that matches your own. Or skim them all."***

As you try the tips and activities, remember:

- Parents do not have to know advanced math to help their child. In fact, parents who are open to it may find themselves learning from and alongside their child. That's positive for everyone.
- The math-positive support of teachers can foster an optimistic math attitude in students—an outcome with long-lasting implications for learning mathematics to a high level.

I am not a perfect mom, a perfect teacher, or a perfect mathematician. If my mistakes in the classroom and at home were lost socks, they would fill a jumbo capacity washing machine on an endless, sudsy loop of regret. But challenges are how we learn, and how we respond to mistakes defines them as shameful or valuable. I hope this resource's math-positive mindset is as contagious as a spring break stomach bug to parents, teachers, caregivers, and eventually to kids. Because, what we adults do and say matters. Our children, whom we love and care for, will emulate our math-positive beliefs and actions. And those kiddos are worth our incalculable effort.

Develop Your Math-Positive Mindset

STEP I

To begin your evolution from math panicked to math positive, you should first explore what a math-positive mindset is and how to cultivate it.

I Can Do It! Math Is for Everyone

CHAPTER 1

1. Don't you have to be "born" good at math? I mean, math isn't for everybody, right?

Too many people believe that mathematics achievement is an exclusive club meant only for those lucky few born with genius genes. Everyone can be good at math with effort. Think of something you do well. Were you good at it from infancy or did you become proficient through effort, practice, and desire? If your children or students idolize an athlete, musician, actor, or public figure, point to the level of effort that person puts into their craft. Hard things are worth the effort it takes to succeed.

Changing how you think about math will change how well you do math. Research shows that people learn best with a growth rather than a fixed mindset (Dweck 2006). In a *fixed mindset*, people believe that they can't improve basic qualities like their intelligence or ability to do well in math beyond an inborn capacity. In a *growth mindset*, people believe that through work and effort, they can improve their abilities to learn and understand. With math, a fixed mindset leads to hopelessness. The growth mindset perspective on learning math is *essential* for mathematics achievement.

"I HAD MY DOCTOR DO A D.N.A. BLOOD ANALYSIS. AS I SUSPECTED, I'M MISSING THE MATH GENE."

Our family has a motto: *I can do hard things*. "Hard things" might be completing math homework, lugging a full trash can to the curb, or folding a fitted sheet. Doing hard things is, well, hard. But that doesn't mean it isn't worth the effort. I even ordered pencils with the *I can do hard things* motto printed on them. We put those empowering pencils to use during homework time as a reminder that effort is more important than natural talent for learning math and for many other things in life.

So, replace the phrase, "I wasn't born to be good at math" with "Math doesn't always come easily to me, but I can work to understand it. And when I do understand it, I can feel good about my effort!"

"PAUSE"ATIVE BOX

Jo Boaler on Growth Mindset

Take a moment to pause and watch Jo Boaler's TEDx-Stanford video, "How You Can Be Good at Math, and Other Surprising Facts about Learning." In this video, Boaler, a professor of mathematics education at Stanford, shares what brain research about the growth mindset teaches us about learning and success in math. Wait 'til you hear what she says about mistakes in math! The findings of her research may surprise you.

2. I was never very good at math. Can I still help my child?

First of all, let me assure you that you can do a lot to help your child through your participation in your child's learning (Hoover-Dempsey and Sandler 1997). You don't have to be a Nobel Prize–winning researcher or hold a doctorate in applied mathematics to help your child see the importance, elegance, and value of learning math.

Regardless of your own level of mathematics proficiency, your role as a parent is vital (Kilman 1999). Just as you probably helped your child learn to read, you *can and should* support your child's developing mathematics skills. In fact, research shows that simply encouraging parents to talk about numbers with their toddlers, and providing them with effective ways to do so, may positively impact children's school achievement (Levine et al. 2010). Your influence will count just as much if not more than that of any math teacher. So, set aside your qualms, try the tips in this book, and feel empowered to help your child succeed in math. Doing so will make a world of difference.

Wall of Mistakes

Cultivate a growth mindset in your students' parents by including growth mindset activities at family math nights. Here's one idea: Have parents add to a Wall of Mistakes by writing their foibles on a sticky note and hanging it up to celebrate how errors help our brains grow.

3. I don't really like math. Does it show?

Uh, yeah, it does show. So, until you catch the math bug, which I hope this book helps you to do, fake it. A child can sense math frustration like a wild animal senses fear. But don't blame yourself too much for your distaste for math. Your negative feelings may not be entirely your fault. Sadly, you may be part of a generation that believed that only some students could learn mathematics. Many students, especially those who were poor, nonnative speakers of English, disabled, female, or members of racial-minority groups, became victims of low expectations for math learning. Students who "struggled" to learn math were tracked into lower-level math classes and given lower expectations for success. Thankfully, today a vision of *mathematics for all* guides modern education. You would not expect your child not to read; similarly, you should not expect your child not to do mathematics.

You may have to call upon the acting skills you honed portraying Puck in your high school's production of *A Midsummer Night's Dream* to shift your own mindset about math homework. You need to transform your inner dialogue from, "Ugh, I hate math homework. It's so hard to know how to help my child" to "Hey, with time and effort I can improve in helping children with their math homework. I can see that they are starting to catch on, and so am I."

Memes about Math Anxiety

TEACHING TIP

Having a sense of humor about math anxiety and a growth mindset can go a long way in winning parents to your cause. Try this idea: Have a contest where parents create original memes using photos of their child. Several free meme generators are available online or as apps.

Parents upload their child's photo and add a witty growth mindset phrase. Parents can email the memes to the teacher for sharing on the class social media page or newsletter. You might even have a contest; the meme that receives the most likes wins a math manipulative to use at home. Here are a few memes I've created using photos of my children:

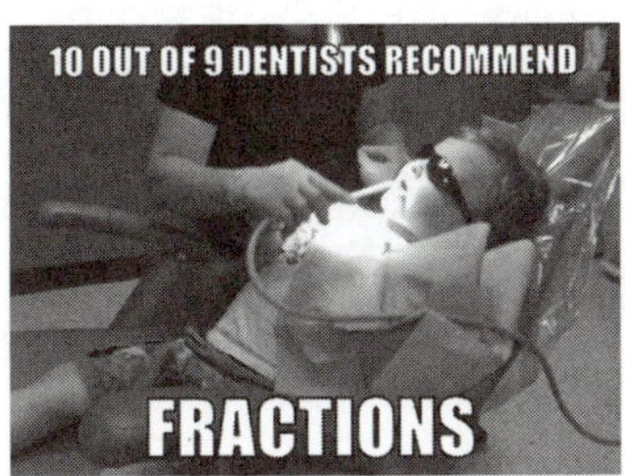

4. How does my attitude toward math affect my child's attitude?

Children get many of their cues about behaviors and values from the adults who care for them (Copple and Bredekamp 2009). I'm sure you've noticed your child mimicking your behaviors, preferences, and mannerisms. Our son Chas holds his fork just like his dad does. McGregor chooses chocolate over fruit just like I do. Recently, Duncan asked for help downloading music onto his cell phone. What music did he want? The '80s alternative music his dad prefers. Children watch adults and do as adults do.

Modeling is an instructional practice where children can observe teachers' thought processes as they demonstrate how to do a problem. Modeling is an essential element of learning, whether that modeling takes the form of showing a child a strategy for solving a problem or talking positively about math. Therefore, *your* attitude about mathematics can influence *your child's* attitude, for better or for worse.

I'll never forget the jaw-dropping comment made by my younger sister's sixth-grade teacher at parent orientation. "I don't like math very much, so we won't be doing much of it this year." How many children missed enriching mathematical experiences because one adult disliked math? This terrible mindset is damaging in countless ways. Adults' math talk must be positive, encouraging, and connected to effort and understanding rather than natural talent.

If you agree that a negative attitude about something forms a stumbling block to success, make a concerted effort to only speak positively about math and math homework. See what a difference a positive outlook can have on a general feeling of self-efficacy for learning and doing mathematics.

5. What is math anxiety?

Math anxiety is a negative emotional reaction or uneasiness (some might call it dread or panic) when asked to complete mathematical tasks (Smith 2013). Math anxiety affects about 50 percent of the U.S. population (Boaler 2012). I'm not sure why there isn't a term for nervousness about other school subjects—maybe *conjugating verbs anxiety* doesn't have quite the lyrical quality that *math anxiety* does. Nevertheless, math anxiety is a complex and real emotional reaction for many adults and children.

Maybe you had an unpleasant experience early on in math that impeded your subsequent learning. Maybe you were surrounded by people who were afraid of math. Maybe you felt competent as a math student, but it's just been so darn long that you're afraid you've forgotten how to do it. Maybe your eyes spin when a colleague casually mentions common denominators and common multiples, and you spend the rest of your lunch break trying to remember the difference between the two terms. (OK, we all know this scenario is ridiculous; no one ever mentions denominators in any context other than math class.) Math anxiety is real, but you can keep it in check during homework time.

6. What can I do to get rid of my math anxiety?

For various reasons, you may be nervous or unsure of your own ability in math and thus your ability to help your child. With math, though, knowledge really is power. The more you learn about math, the less scary it becomes.

A few months ago, I spoke with a mother of two preteens who confided that she never understood why a square number was called a "square number." When she helped her fifth-grade son complete a homework activity that used hands-on activities to explore the topic using a grid, my friend had an aha moment. Four, nine, sixteen . . . they all make a square when you arrange them in a grid format.

1	2
3	4

1	2	3
4	5	6
7	8	9

1	2	3	4
5	6	7	8
9	10	11	12
13	14	15	16

Square numbers using grid paper

If you fear learning more about math, don't let anxiety slow you down. You might want to try an experiment I use in my Math for Elementary School Teachers courses. My students are in their final few semesters before they begin their teaching careers. On the first day of class, I have them write on strips of paper all their misgivings, anxiety-laden thoughts, and personal doubts about themselves as doers of math. We put the strips in an empty jar and set them on the shelf. While we can't destroy them permanently (like by setting them on fire, though it's tempting), we can set them aside so that they don't get in the way of our learning. I ask my students to promise me they will never disparage math in front of their students (or even privately, since I believe it feeds feelings of negativity). We then try to approach the course as eager learners and mathematically powerful thinkers.

Try the same experiment. Even if you don't make that symbolic gesture to fill a jar with your doubts, remember that it won't help your child to hear you say that you were never good at math, you could never understand math, or you never liked math. Don't send that message! Math-positive communication makes for math-positive thinking. Focus on what you *can* offer your child: support, encouragement, and high expectations for their success.

"PAUSE"ATIVE BOX

Adults' Math Anxiety

Following are comments from teachers and parents that I've collected over the years. Take a moment to pause and think about these comments; can you relate to some of them? What ideas might you have for alleviating math anxiety in adults?

- "I would like to help him with homework, but I get impatient and that makes us both feel bad." —B. S., dad of one
- "My children laugh at me when I offer to help them with their math homework! Past sixth grade, I'm useless to them!" —S. D., mom of four
- "When I was offered a job teaching math instead of language arts, I was genuinely freaked out. I struggled with math in school and never felt like I got it. I hope I can switch to a reading classroom next year." —M. A., elementary school teacher
- "I made a deal with my partner that he would be in charge of all math help. I have always been afraid that I would teach [my kids] my math anxieties and insecurities." —L. J., mom of four
- "Can't we just do everything with a calculator?! And that's me saying that, not my kids!" —E. C., mom of three
- "I'm a nanny, not a math expert. When the kids I take care of need help with math, I have them look it up on the internet." —T. J., nanny
- "I am not exaggerating when I say that my brain shuts off where math homework is involved. Right now, my twin girls are only in kindergarten, so I can handle helping them, but I am horrified to think that I will have to assist in stuff I don't understand. Guess I better save my money for a tutor." —C. B., mom of three
- "The way they teach math is so different from how I learned it. I don't want to mess my daughter up by telling her how to do it differently from the way her teacher does." —S. B., dad of one
- "The kids don't even bring home a textbook I can sneak a peek at for a refresher. I used to love math, but it's been a long time." —C. C., grandfather of four
- "Today I reviewed one of Luke's eighth-grade math tests and literally felt my right eye twitching and my heart racing. Anxiety does not begin to describe it!" —S. L., mom of two

7. How do I ease my child's math anxiety?

In third grade, my son Duncan had to pass a "Monster Test" to show he had memorized the times tables. He had to solve something like 2,478 multiplication problems in fifty-three seconds—or so it felt to him. He agonized over the test for weeks, marking it on the family calendar for all to dread. I contemplated the best approach for easing his anxiety about this timed test (more about my own distaste for timed tests in Chapter 2). The following tips helped Duncan to keep focused on effort and optimism, and they may help your child to feel less stressed during math class or homework time.

TEACHING TIP

Class Techniques for Easing Math Anxiety

Try these simple relaxation techniques to ease students' anxiety before or during assessment.

- *Helicopter Breathing:* Inhale deeply. Expel breath while making a *ch-ch-ch* sound imitating a helicopter. Swirl both index fingers in a circular motion while raising them above the head. When arms are fully extended, blow out the rest of the breath with a straight *sh* sound and lower the arms. Repeat until calm.
- *Body Part Relaxation:* Focus on relaxing individual muscles one by one. In a calm, slow voice the leader says, "Focus on your eyebrows. Raise them several times then relax them. Focus on your jaw muscles. Bite down hard then relax. Move next to the shoulders. Raise them and squeeze your shoulders together in the back. Then roll them slowly and relax them." The leader continues down the body until even the toes are relaxed.
- *Singing Bowl:* Strike a singing bowl and have students focus only on the sound of the bowl ringing. The note may be heard for up to a minute if students listen carefully. The tone of the bowl has been found to induce a peaceful state and clarity of mind while engaging the relaxation reflex and inhibiting stress.

"PAUSE"ATIVE BOX

Easing Children's Math Anxiety

Following are ideas I've found useful in easing children's math anxiety. Take a moment to pause and think about these suggestions. Which ones would you like to try? Might some of these ideas help adults as well with their math anxiety? What additional ideas might you have?

- Help children understand they are not alone in feeling nervous when they don't understand how to do something. Tell them of a time when you encountered something challenging, worked hard to understand it, and overcame your anxiety.
- Practice stress management techniques such as relaxed breathing and positive self-talk. Children may want to use these techniques during testing situations as well as homework time.
- Encourage children to ask questions when they don't understand something. Some people think asking questions is a sign of weakness; however, explain to children that asking questions makes their brains work better. The brain makes connections called neural pathways between previous learning and new information. We need neural pathways to make sense of new ideas.
- Support children in doing math in a way that's natural for them. There's often more than one way to work a math problem. Help children find what math strategies work best for them.
- Break math assignments into manageable parts. Sometimes it's easiest to divide and conquer. If a homework assignment is multiple pages long, take a break between sections.
- Do the easiest sections of a test or homework first. Children will start the homework session feeling like a success, which will help them relax during the tougher portions.
- During homework time, keep the atmosphere upbeat and support children with positive reinforcement and encouragement for effort.

8. I want to be supportive of my child, but I am not a fan of praise since it seems to turn my child into a bit of an "approval junkie." How can I build my child's confidence in appropriate ways?

Once upon a time, parents were told they couldn't cheer loud enough for their children. "You're amazing!" became "You're super amazing!" Then researchers like Alfie Kohn (2001) discovered that some children became "approval junkies" who thrived on parental approval to such a degree that they felt like failures if they didn't receive praise for every attempt. Keeping praise at the same level as a child's effort and accomplishment builds confidence (Henderlong and Lepper 2002). Whatever your feelings about praise versus encouragement, most parents would agree that children need to know they are loved and supported regardless of whether they struggle or soar. For more on the subject of praise, see Chapter 3, Question No. 13, *How can I help my child to persevere when math is challenging?*

"PAUSE"ATIVE BOX

Praising Effort

Take a moment to pause and think about encouraging comments you can share with children that acknowledge their effort. Here's a list of my favorites for starters.

- I see how you used a strategy to solve the problem.
- I like how you kept trying even when it was tough.
- I think your teacher will appreciate your effort.
- Thank you for starting your homework without being reminded.
- It really shows when you make your best effort.
- You should feel proud for working until the job is done.

Giving Feedback on Effort

TEACHING TIP

Focusing on students' efforts rather than their talent for mathematics is important for building a growth mindset. The following are some of my favorite ways to give feedback on effort. How might focusing on effort help build children's understanding of the connection between work and success?

- I can see you took your time to do your best.
- I bet that makes you feel proud that your hard work paid off.
- You showed your work even though it was a lot of writing.
- You didn't give up on some tough problems.
- You didn't let mistakes fluster you.
- When the problems got harder, you worked harder.

9. My daughter thinks math is just for boys. How do I handle this ridiculous notion?

Just like there is no math DNA, there is no math chromosome. If more girls were given the same support and opportunities as boys to excel at mathematics, perhaps there would be many more high-achieving girls and women in mathematics. Unfortunately, even today, young girls may not be encouraged to investigate the world in the same way that boys are. Well-intentioned parents may still primarily give girls toys that encourage sociodramatic play such as dolls, kitchen sets, and so forth. When Mattel Toys created the Teen Talk Barbie doll in 1992, one of her preprogrammed statements said, "Math class is tough" (Sullivan 1992). Barbie's words emphasized the gender bias that portrays girls as less skilled at math. Conversely, boys may be given blocks, chemistry kits, and construction tools and encouraged to explore their world in a more mathematical way than girls are.

Researchers have found that, because most caregivers of young children are female, this stereotype may affect a child's earliest math experiences (Beilock et al. 2010). Caregivers need to have high expectations for both genders and encourage all children to follow their mathematical curiosity, strive to make sense of mathematical ideas, and work to excel in mathematics. Parents must notice, value, and build on children's excitement as they explore mathematics from the earliest ages (Eisenhauer and Feikes 2009).

Margot Lee Shetterly's book *Hidden Figures: The American Dream and the Untold Story of the Black Women Mathematicians Who Helped Win the Space Race* (2016) draws attention to the contributions of black females employed, due to their speed and efficiency, to compute equations used in aeronautical engineering during the early days of the United States' space race. The book, also adapted as a picture book and a young readers' edition, exemplifies what women can achieve in mathematics through hard work and perseverance. Several books by actress-turned-author Danica McKellar (remember the adorable Winnie on *Wonder Years*?) offer girls a vision of themselves as mathematicians. McKellar, who holds a degree in mathematics, writes picture books and books for middle school– and high school–age readers. Your librarian or bookstore can direct you to many other inspirational biographies and motivational titles highlighting females' achievements and contributions to the field of mathematics.

10. I really like math. Why doesn't my child seem to feel the same way about it?

While some parents struggle to overcome their negative feelings about math, others love it and can't see why their kids don't love it as well. Often, we as parents try to map our own talents and preferences onto our kids. My husband and I loved music when we were teenagers. I sang in choirs and at festivals and my husband played bass guitar in a band called Yellow Snow. We couldn't get enough of performing and writing music. But our children have different interests. Duncan, for example, grudgingly took piano and cello lessons but never caught the passion. Music doesn't speak to him the way coding, comics, and concocting stuff in the kitchen do. Despite our efforts to coax our kids to follow our passions, they must discover their own. So, if you find yourself trying to mold your kid into a mini-me, consider the great opportunity you have to let children be their own best self. You love math. Great. Continue to love it and talk positively about it. Also be understanding of your child's growth into loving it too. It's a journey.

11. What books can I share with my children to show how math-positive mindsets can be applied in many different situations?

Children's book authors are starting to catch on that persistence, problem solving, and positivity are powerful words in mathematics and in life. Here are a few books you can read with your child to see how math-positive mindsets can be a force for good.

Book	Summary
Hallowell, Edward. 2004. *A Walk in the Rain with a Brain*, illus. Bill Mayer. New York: Harper Collins.	Fred, a brain who has lost his head, explains that "whatever you do at great length, your brain will grow there in strength." Encourages play as a form of learning.
Deak, JoAnn. 2010. *Your Fantastic Elastic Brain: Stretch It, Shape It*, illus. Sarah Ackerley. San Francisco: Little Pickle Press.	This colorful book teaches the parts and functions of the brain. Emphasizes how the first ten years of life are ideal for building brain strength and neural connections as children work to learn many different things.
Pett, Mark, and Gary Rubinstein. 2011. *The Girl Who Never Made Mistakes*, illus. Mark Pett. Naperville, IL: Sourcebooks Jabberwocky.	Most people in town only know Beatrice as "The Girl Who Never Makes Mistakes." But is she missing out by not being willing to take risks and try new things? When Beatrice messes up during a talent show, she learns to laugh at herself and her mistakes.

(continued)

(continued)

Book	Summary
Elliott, David. 2015. *Nobody's Perfect*, illus. Sam Zuppardi. Somerville, MA: Candlewick.	This book explores ways we can find joy and fun in not being perfect. From messy rooms to loud little sisters, coming *close* to perfect is enough for this little boy.
Reiley, Carol E. 2015. *Making a Splash*, illus. Jason Pastrana. San Francisco: Go Brain!	This book directly teaches growth mindset principles through a story of two siblings learning to swim.
Yamada, Kobi. 2016. *What Do You Do with a Problem?* illus. Mae Besom. Seattle: Compendium Inc.	A problem appears as a cloud over a boy's head. He worries a lot about his problem, which makes the problem swell. Finally, the boy attacks the problem and learns it has something beautiful inside—an opportunity to learn, grow, and be brave.
Saltzbeg, Barney. 2015. *Beautiful Oops*. New York: Workman Publishing.	This engineered book shows how a torn piece of paper or spilled paint can have lots of possibilities. It teaches that mistakes are worth exploring. Search online for a video of Saltzberg showing school children how mistakes can be adventures in creativity.
Cox, Lisa. 2017. *Not Yet*, illus. Lori Hockema. Indianapolis: Dog Ear Publishing.	Lorisa keeps trying despite making lots of mistakes. "Will she make it? You bet! Is she there? Not yet." This book encourages a growth mindset in the face of challenges.
Bardoe, Cheryl. 2018. *Nothing Stopped Sophie: The Story of Unshakable Mathematician Sophie Germain*, illus. Barbara McClintock. New York: Little, Brown Books for Young Readers.	This book is an inspiring biography of the first woman to win the Royal Academy of Sciences grand prize. Despite having to publish under a male pen name, Sophie Germain's contribution to mathematics was the result of persistence and drive.

For more on literature to support children's math learning, see Chapter 3, Question No. 26, *My children love to read. I'm wondering if I can get them excited about math through reading. What are some good picture books that reinforce some of the math concepts my children are learning at school?*

QUICK REFERENCE CHART

Common Concerns about Math Attitudes

This table is a quick reference for some of the *most common concerns* related to math attitudes. The table lists the concern, an explanation of the thinking behind it, and an idea that addresses the concern. It really is true that our attitude determines our effort, which in turn affects our outcomes. Work to improve your math-positive mindset by trying out the suggestions in the *What to Do* column.

Concern	Explanation	What to Do	Question Number(s) (to Learn More)
Math myths (you must be born good at math)	This is a lingering misconception. Brain research shows there is no such thing as a math gene.	Go online and find Carol Dweck's TEDx Talk entitled "The Power of Yet" and watch it with your children. Discuss how the brain of a person with a growth mindset responds to challenges and to mistakes.	1, 2
I'm not good at math.	Since there is no inheritable math gene, don't let past negative experiences with math prevent you from being a math-positive parent.	Math may look a bit different than it did when you were in school. Learning new ways of understanding math *alongside your children* may give you exciting new insights into math. It's okay to learn together.	2
Overcoming negative past experiences with math	You may not have had the benefit of learning math using the hands-on, problem-rich methods found in today's classrooms. Give math a second chance! You may just find you like it this time around.	Training your brain to think positively about math is the first step. Then, work to ensure your math talk with children is encouraging and connected to effort and understanding rather than natural talent.	3, 4
Math anxiety	Math anxiety is a genuine psychological phenomenon. But it doesn't have to define your relationship with math or with children.	Math-positive communication makes for math-positive thinking. Focus on what you *can* offer your children: support, encouragement, and high expectations for their success.	5, 6, 7

(continued)

(continued)

Concern	Explanation	What to Do	Question Number(s) (to Learn More)
Praise versus encouragement	Praise focuses on products. Encouragement focuses on processes.	Try these encouraging comments: • "I can see you're sticking to it even though this is a hard problem." • "You show a lot of persistence." • "Your teacher will be impressed that you used so many different strategies."	8
Boys are better at math	Just like there is no math DNA, there is no math chromosome. Maintain high expectations for both genders. Encourage all children to follow their mathematical curiosity, strive to make sense of mathematical ideas, and work to excel in mathematics.	Share books with math-positive female characters. Talk about women's contributions to mathematics and science. Emphasize that competence in mathematics opens doors of opportunity for all children.	9
How can I help my kids enjoy math as much as I do?	If you find yourself trying to mold your kid into a mini-me, consider the great opportunity you have to let children be their own best self.	Continue to love math and talk positively about it. Also be understanding of your child's growth into loving it too. Finding how your children's interests overlap with math will engage them more in authentic ways.	10
Books for growing a math-positive mindset	From celebrating mistakes to taking chances, many picture books show kid-friendly examples of math-positive mindsets.	Check out the suggestions provided and explore new ones with your children.	11

Get to Know Your Math-Positive Team

STEP II

The second step in building a math-positive mindset is to learn more about your math-positive team. For both the teacher and parent, this means getting to really know the math classroom—and getting to know each other!

Chapter 2 gives parents a glimpse into the secret world of teachers and their classrooms. Okay, it's not quite as dramatic as that. But it's helpful for parents to understand why teachers do what they do. The chapter opens with suggestions on how to have successful parent-teacher conferences and how to word emails to teachers (and back to parents) without freaking them out. The chapter then moves into understanding why your child's math classroom might look and feel different from the classrooms you remember as a child. Among other things, this chapter explains why teachers emphasize problem solving (because we want children to *think* not just follow procedures) and why we ask children to explain their answers (because we're interested in their flexibility in math reasoning). For teachers, the chapter provides helpful tips for using sentence frames to guide students' reasoning and communication in math and alternatives to timed tests for assessing fact fluency. The chapter concludes with a couple of Q and A's on state standards.

I Can Get to Know My Child's Math Teacher and Classroom

CHAPTER 2

(continued)

(continued)

TEACHING TIPS

1. I'm sort of intimidated by my child's teacher. How can I overcome this?

Remember that teachers are real people. They strive to help your child succeed and are looking for your help. You and your child's teacher are on the same side, working together to give your child a firm foundation in mathematics. Since you are in this together, it's a good ideas to communicate often and positively.

It's OK to ask your child's teacher questions. If you don't understand how or why, ask. However, ask nicely. I've found that a quick email or note with a casual tone works best. For example, "Hi, Mrs. Cannon. Just a heads up that Duncan is really confused about subtracting two-digit numbers, especially when he has to regroup. We worked on it for thirty minutes until we both wanted to pull our hair out. Help! I'd love any ideas you can offer."

Also, let your child's teacher know when you've found an activity that seems to help your child feel positive about math. For example, "Hi, Mrs. Cannon. I just had to share our success from last night's homework. At first, Duncan seemed a bit confused about finding common denominators for fractions. It seemed to help when I called them equivalent fractions. Then lightbulbs really starting going off when we got out the measuring cups and Duncan used the $\frac{1}{4}$ cup to fill the $\frac{1}{2}$ cup. Ding! Ding! Ding! I'm doing my happy dance! I'd love any other ideas you might have for making this concept make sense to Duncan."

2. Why is it important to have open lines of communication with my child's teacher?

Some parents (and teachers) wait until a problem arises to engage with one another. However, I encourage you to begin *now* to cultivate a partnership with your child's teacher so that when challenges arise, the two of you can spring into coordinated action.

"PAUSE"ATIVE BOX

Ideas for Building a Relationship with Your Child's Teacher

Take a moment to pause and think about the following ideas for building a relationship of trust with your child's teacher. What other ideas might you suggest?

- Send a quick note or email when things are going well.
- Offer appreciation for extra effort you see the teacher make on your child's behalf.
- Gratitude does not have to be tied to gifts of any sort. A short email to the principal or school director praising your child's teacher lasts much longer than a bottle of smelly lotion or candy bar.
- Volunteer when possible to lighten the teacher's load.

3. What is the best approach for parent-teacher conferences with my child's teacher?

The primary purpose of the conference is to provide you with information about your child's academic progress. But it's also a great opportunity to build trust in the parent-teacher partnership, brainstorm ways you can support your child's future learning goals, share concerns and opinions about school, and address those concerns (Constantino 2016). I recommend beginning the conference by letting teachers know that you appreciate and support their efforts. If a teacher asks if you have questions, by all means have some ready.

Reaching Out to Parents

TEACHING TIP

Don't wait until a student is sinking to reach out to parents. Building a relationship of trust takes time, so start sooner rather than later to get parents on your side of the math learning equation. Consider the following:

- *Positive Notes:* Send students home with positive notes; showing genuine care for your students goes a long way to building relationships with parents.
- *Newsletter:* Send a newsy newsletter from time to time; keep it brief and free of grammatical and punctuation errors.
- *Back-and-Forth Books:* A simple composition book works fine for a back-and-forth book. Be sure to balance good and bad news, use everyday language, and plan a schedule for how often the book should make the trip home and back. These books are especially helpful for children with special needs.
- *Reminders:* The school website or voice mail system as well as commercial text message apps can be used to highlight a homework due date or send a reminder about an upcoming event.
- *Emails:* Emails can easily be blasted to whole classes or personalized for a student.
- *Phone Calls:* Phone calls don't have to be saved for bad news. Positive phone calls can make a parent's whole day, week, even month! If you have a hunch a child may be giving you a run for your money this year, start off with a few positive calls early on. That way, when the trouble comes to a head, you have a parent partner who knows you, knows you care about their child, and knows you are reaching out with genuine concern.

"PAUSE"ATIVE BOX

Questions for Your Child's Teacher

Following are a few general questions I like to ask my children's teachers. Take a moment to pause and think about these questions in relation to your child. Which ones might be most helpful for you? What other questions might you have?

- What are my child's strengths?
- What are some successes you have seen in my child this year?
- What areas does my child need to work on?
- Is any tutoring available before, during, or after school?
- Do you offer help on a one-on-one basis or in a group setting? When?
- What tutors might you recommend outside of school?
- Where can we find more problems to use for practice?
- What are some mathematics-related websites we can visit for extra practice?
- What programs does the school offer for remediation or enrichment?
- What can I do to help you?

TEACHING TIP

Ensuring Successful Parent-Teacher Conferences

Parent-teacher conferences can be a great way to support students' learning. Following is a list of considerations when planning conferences. What other ideas might you have for planning and carrying out math-positive parent conferences?

- *Format:* Conferences can be face-to-face but also via email, phone, or written note.
- *Language:* Make sure all parents feel invited and encouraged to attend; for parents who may not be as comfortable speaking English, offer translation services.
- *Schedules:* The time and location may be flexible based on parents' schedules. I've even met a parent at Starbucks for a quick portfolio handoff when our schedules couldn't match up any other way.

(continued)

(continued)

- *Seating:* If you are meeting face-to-face, plan to be seated beside the parent rather than sitting at your desk and relegating the parent to a child-sized chair. The latter may send the message that you are not interested in partnering with parents and view yourself as The Boss. Sitting beside the parent as you share the child's work allows you both to view and discuss the materials together as a team.
- *Student Work:* Plan ahead what you'll discuss and select samples of the student's work to share. You will want to include checklists, photos, anecdotal records, and numerical grades that show the child's progress. You may want to put sticky notes on the work samples to guide the conversation and to help you focus on what is working for the child. Avoid jargon and be sure to let parents offer ideas and ask questions about their child's progress and how they can support learning.
- *Communication:* Be positive; offer a few suggestions or goals for addressing needs and express appreciation to the parent for taking time to attend the conference. When parents ask questions for which you are unprepared, it's OK to say that you don't know but you'll find out. Don't invent an answer on the spot. Tell parents that you need additional time to think about a response. Then do the research required and get back to them as soon as possible. If you must address serious concerns you have about a child's progress, be clear and honest but also empathetic. If students would most likely benefit from additional one-on-one attention, you'll likely need to enlist their parents' help in finding a solution. If you have a relationship of trust with parents, the process will be much smoother. Communicating with parents requires effort and time, but it is worth it to build a successful home-school partnership.

4. How can I better understand what's going on in the math classroom?

Reading this book is a good start to helping you understand how math is taught today. If possible, you might also make a visit to your child's classroom. This will allow you to physically handle the materials children are using, hear the language of mathematics, and observe how children work collaboratively to solve problems and build understanding. Your child's teacher might ask you to help students play a math game or join in a learning station. At an appropriate time, you can ask the teacher how these activities and routines help children learn mathematics. The teacher might also share children's work to demonstrate certain points. Remember, when you're setting up a time to visit, approach the teacher with genuine interest in the exciting changes in math instruction, not with a skeptical attitude about the curriculum or teaching approaches. Tell the teacher you want to soak up some math understanding so that you are better able to help your child succeed in math. You may leave your visit feeling reassured that the instructional practices and expectations the teacher has for your child are building a math-positive mindset. And don't be surprised if you feel yourself becoming a more math-positive parent at the same time.

5. Why is math these days so different from how I learned it?

Your child's mathematics classroom may look and sound quite different from how you remember your own student days. Instead of straight rows of desks, you may see children scattered around the room working in small groups. Instead of the teacher demonstrating a problem on the board and students silently copying the steps, you may see students working together to investigate a problem while the teacher serves as a facilitator. Instead of students reciting memorized rules and

Engaging Parents in Children's Learning

TEACHING TIP

Don't feel intimidated by parents who are interested in their child's learning. Engaged parents are powerful partners in the education process. If a parent contacts you with questions about curriculum or teaching approaches, check with your administration to see if parents are welcome in the classroom during math time. Chances are, administration will leap at the chance to showcase your all-star approaches to a math-positive mindset. Invite the parent to join the class and be sure to design a lesson that offers a glimpse of the hands-on, engaging approaches you know are most effective for learning concepts deeply. Seeing and hearing a class full of children making sense of math is the best way to help parents buy in to today's approaches. And, we all know that powerful partnerships between home and school positively affect learning outcomes for children. Be yourself. Be true to the mathematics you love. And wow 'em with the best practices you use every day for building math-positive mindsets in your classroom.

procedures, you may see students discussing and questioning solutions to problems. Children may even be part of a "flipped classroom," where they watch a video online the night before math class and then solidify concepts introduced in the video the next day in class with the teacher and other students.

If changes in math education have you confused, you aren't alone. Lots of parents have questions about reforms (Hendrickson et al. 2004). Sometimes these changes can make a parent feel a bit "out of it." It's important to recognize that, over the past few decades, educators have greatly improved their understanding of how students learn mathematics (Mirra 2004). We know that learning is not a passive activity but an active process of building new knowledge from experience and prior understanding. With this realization, we must use new approaches that utilize communication, connections, representation, problem solving, and reasoning. That's why the math classroom, assignments, and strategies are so different from how you and I learned them—primarily through drill and practice and memorization. Learning to think, persist in the face of failure, and adapt to change will help prepare today's students to enter a workforce in which 85 percent of jobs that will exist in 2030 have not yet been invented yet (Institute for the Future and Dell Technologies 2017).

6. My child's teacher talks about integrating math with other subjects. What does this mean?

As a parent, you are a master at integration and being as efficient as possible with your time and resources. You might integrate learning and exercise by listening to a podcast while working out. I often integrate playing with my kids and cleaning the house by singing silly songs and chasing them around with a feather duster "tickler."

At school, children's mathematics experiences may be woven into the other curriculum areas, including science, social studies, fine arts, physical education, and language arts as appropriate. Children might integrate math and art by composing shapes that create a Picasso-inspired self-portrait or integrate math and science by analyzing nutrition labels on food containers. Brain research shows that long-term memory, or true learning, depends on information that makes sense and has meaning. Integrating math with other subjects helps children build understanding by connecting new information with previous or concurrent learning. Without these connections, children's learning experiences would add up to a collection of miscellaneous topics and unrelated facts (Sousa 2007).

One of the goals for your child's math teacher is to select rich, integrated mathematical tasks and problems that are accessible for all children, yet challenging enough to help them grow in mathematical understanding (Hiebert et al. 1997). Sometimes these rich problems will be extended—even to the home. For example, in elementary school Duncan conducted an integrated mathematics and social studies project that involved researching and selecting a gift for a friend. He selected a toy, gathered information about its cost and availability online and in stores, came up with a plan for earning the money to pay for the gift, and presented his findings in a poster display. His experiment integrated social studies, language arts, technology, art, and mathematics in a real-world context that was personally meaningful to him. And the gift was for me! A stuffed animal with a T-shirt that said, "I ♥ Math." How can you go wrong with that?

Integrated Instruction and Project-Based Learning (PBL)

TEACHING TIP

Integrated instruction is more intentional than killing two birds with one stone. It's looking for meaningful ways to show the interconnectedness of mathematical ideas, connections between mathematics and other subject areas, and the countless essential connections of mathematics to the real world. It is a powerful way to build a math-positive mindset in students because it shows the relevance of math in all aspects of life.

Project-based learning (PBL) is a popular, effective way to integrate mathematics with other subjects and with students' interests. In PBL, you pose to students a driving question that engages them in investigating a problem or process. The driving question should be open-ended and objective. It might begin with "How can I" followed by a challenge and an audience. For example, *How can I plan a family math night that parents and kids will want to attend?* or *How can I plan a field trip the school board will approve?* The students read books, interview people, apply mathematics, and write up a plan they present to their audience. PBL is just one approach to integration but has been shown to increase real-world applications of mathematical ideas. Whatever conduit you use to connect math to students and students to math, the authenticity of the experience is sure to tie together theory and practice in ways that are meaningful to kids.

7. Why is learning math concepts with understanding so important? Isn't it enough to just be good at following steps or procedures?

Mathematical procedures are like recipes developed by efficiency experts to enable people to go straight to specific approaches to particular kinds of problems (Schwartz 2008). Sometimes we call these procedures algorithms. Algorithms are quick, efficient approaches to doing computation. You might have learned this cute little rhyme in middle school, "Ours is not to question why, just invert and multiply." This ditty was supposed to prompt you to invert the second fraction when dividing a fraction by a fraction or a fraction by a whole number. But did memorizing the rhyme and applying it on worksheet after worksheet help you to understand the mathematics behind the procedure? Probably not. Algorithms and other procedures aren't necessarily intuitive or logical, especially if you don't understand what's really going on in the process. What if we thought of cooking a delicious dish as nothing more than following another person's recipe? That would take a lot of the fun, creativity, personal investment, and learning out of the cooking experience. And what if we forgot or couldn't find the recipe we needed? Could we still replicate a dish? We could if we knew the ingredients and the fundamentals for cooking. In today's mathematics classrooms we are teaching concepts first and foremost so that children can understand the underlying processes and apply them in lots of different contexts. Though learning math conceptually may take longer than memorizing steps or procedures, the learning will be more robust, less fragile, and more easily transferred to new contexts. It builds math-positive mindsets, and it's worth the effort!

"PAUSE"ATIVE BOX

Benefits of Learning Mathematics Conceptually

Learning mathematics conceptually has some real benefits for children. Take a moment to pause and think about the benefits; I've listed a few of them here. What other benefits might there be?

- Children are less likely to forget concepts than procedures. If children have conceptual understanding, then they can reconstruct a procedure that may have been forgotten. On the other hand, if procedural knowledge is the limit of children's learning, there is no way to reconstruct a forgotten procedure.
- Children can apply conceptual understanding to new types of math problems.
- Children can apply conceptual understanding to real-world problems—one of our overarching goals as parents and teachers.
- Research shows that students in a conceptually-oriented mathematics class outperform students in a procedurally-oriented mathematics class on tests and on measures of attitude toward mathematics (Boaler 1998; Cain 2002; Fuson, Carroll, and Drueck 2000; Hiebert and Grouws 2007).

8. My child's teacher mentions mathematical processes. What are these and why are they important?

When you were in elementary school, you probably spent most of your math lessons completing arithmetic problems. Today we understand that mathematics is much richer and more multifaceted. The National Council of Teachers of Mathematics (2000) has stated that the processes behind mathematics learning are as important as the content itself. NCTM's mathematical processes are (a) problem solving, (b) reasoning and proof, (c) communication, (d) connections, and (e) representation. When our math classrooms move beyond lockstep procedures to include messy, interesting, real-world problems, children

TEACHING TIP

Self-Check: Are Your Math Tasks Powerful Enough?

How much of your instruction in mathematics centers on rote procedures and how much revolves around the mathematical processes of reasoning, problem solving, communication, connections, and representation and proof? It depends largely on the tasks you provide for students. Here are eight criteria you can use for self-reflection about the tasks you present to students.

1. Are the tasks you choose to use **essential**, fitting a big idea in mathematics or are the tasks tangential or skill-based?
2. Do students find the tasks **authentic** and value the outcome of the task or do they find the tasks contrived to provide practice with a skill?
3. Do you fill students' time with superficial busy work or **rich** tasks that lead to other problems, raise interesting questions, and have many possibilities?
4. Do the tasks you provide **engage** students, foster persistence, and stimulate their thinking or are the tasks uninteresting?
5. Do the tasks lend themselves to students' **active** decision making and collaboration as students construct their own meaning or do the tasks place students in a passive role?
6. Are the tasks **equitable**, developing a variety of thinking styles and positive attitudes about mathematics, or are the tasks inequitable, appealing to only a narrow student population?
7. Are the tasks **open**, with many right answers or at least many possible avenues for getting a right answer or are they closed, inaccessible to a full range of students?

The words in bold (based on work by Stenmark 1991) are powerful benchmarks for high-quality, process-oriented tasks that focus on conceptual understanding rather than rote procedures. Periodic self-reflection ensures that you stay on target with a body of essential, authentic, rich, engaging, active, equitable, and open tasks for effective, high-quality mathematics experiences.

learn about logic, reasoning, and problem solving—mathematical habits of mind and lifelong abilities that connect math to their everyday lives. These are the substance of math-positive mindsets! When children can accurately and precisely represent their mathematical ideas with words, drawings, and symbols, they have a new communication skill that will serve them well regardless of their chosen career path. For more on developing mathematical habits of mind, see Message 31, "Developing Mathematical Habits of Mind" in Cathy Seeley's book, *Smarter Than We Think* (2014). Seeley's book is written for parents, teachers, and anyone interested in improving mathematics education.

9. There is so much vocabulary to learn in math. Why is the language of mathematics important?

Math time is speaking, reading, and listening time, not just computation time. Where math was once a solitary, silent endeavor, much of mathematics today is language dependent (NCTM 2014). Strong vocabulary increases children's confidence in sharing their mathematical ideas and helps them make connections between math and the real world, represent their ideas, explain their thinking, and justify their responses.

10. I can't keep all these terms straight. Am I the only one struggling?

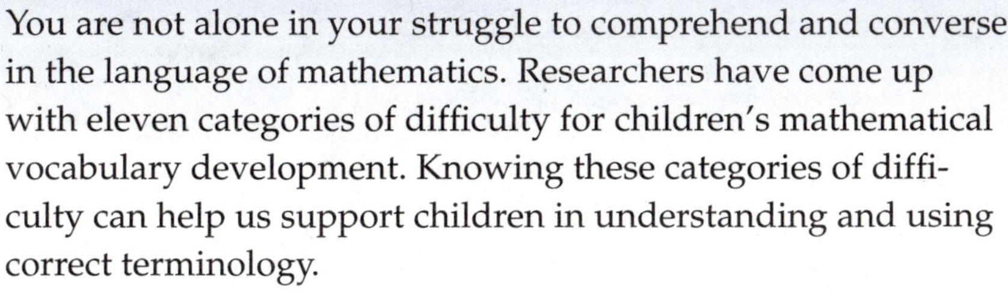

You are not alone in your struggle to comprehend and converse in the language of mathematics. Researchers have come up with eleven categories of difficulty for children's mathematical vocabulary development. Knowing these categories of difficulty can help us support children in understanding and using correct terminology.

"PAUSE"ATIVE BOX

Eleven Categories of Difficulty for Mathematical Vocabulary Development

The following table outlines the categories of difficulty (Rubenstein and Thompson 2002). Take a moment to pause and think about each category. What additional examples might you have for each?

Category	Explanation	Examples
1.	Words may be shared by mathematics and everyday English but have different meanings based on context.	*Right* angle versus *right* answer *Right* angle versus *right* hand *Reflection* as flipping over a line versus *reflection* as thinking about something *Foot* as 12 inches versus the *foot* on a leg
2.	Words shared with everyday English may have comparable meanings, but the math meaning is more precise.	*Difference* as the answer to a subtraction problem versus *difference* as a general comparison *Even* as divisible by 2 versus *even* as smooth
3.	Some math terms are found only in math contexts.	*Quotient*, *decimal*, *denominator*, *quadrilateral*, *parallelogram*, *isosceles*
4.	Some words have more than one mathematical meaning.	*Round* as a circle versus to *round* a number to the tens place *Square* as a shape versus *square* as a number times itself *Second* as a measure of time versus *second* as a location in a set of ordered items *Times* as in multiply and *times* as hours and minutes
5.	Some words shared with other disciplines (content areas) have different meanings based on context.	*Divide* as to separate into parts versus Continental *Divide* which separates eastward and westward flowing waters *Variable* in mathematics is a letter that represents possible numerical values, but *variable* clouds in science are a weather condition.

(continued)

(continued)

Category	Explanation	Examples
6.	Some math terms are homonyms or homophones with everyday English words.	*Sum* versus *some*; *arc* versus *ark*; *pi* versus *pie*, *graphed* versus *graft*; *cent* versus *scent* and *sent*.
7.	Some math words are related, but students may confuse their distinct meanings.	*Factor* and *multiple*, *area* and *perimeter*, *hundred* and *hundredths*, *numerator* and *denominator*
8.	A single English word may translate into Spanish or another language in two different ways.	In Spanish, the *table* at which we eat is a *mesa*, but a mathematical *table* is a *tabla*.
9.	English spelling and usage have many irregularities.	*Four* has a *u*, but *forty* does not. Fraction denominators, such as *sixth*, *fifth*, *fourth*, and *third* are like ordinal numbers, but rather than *second*, the fraction term is *half*.
10.	Some math concepts are verbalized in more than one way.	*Skip-count by threes* versus *tell the multiples of 3* *One-quarter* versus *one-fourth*
11.	Students may adopt an informal term as if it is a mathematical term.	*Diamond* for *rhombus* *Corner* for *vertex*

From "Understanding and Supporting Children's Mathematical Vocabulary Development" by Rheta Rubenstein and Denisse Thompson (NCTM, 2002). Reprinted with permission.

11. Why are children asked to show their work? My children can do it quickly and accurately in their heads. It seems like all that writing is just slowing them down.

I agree that writing down all the steps used to solve a straightforward math problem can be tedious. My son Duncan and I have battled this problem many times. There are many ways to solve problems in math. We might use paper and pencil to record calculations, make lists or tables, or draw pictures. We might use mental math to figure something out "in our heads."

We might use a calculator or some other tool. We might use manipulatives such as blocks, dry beans, or counters. When teachers ask to see students' paper-and-pencil work, they want to "see" what students are thinking as they solved the problem. For more insights on writing in math, see Chapter 3, Question No. 14, *Sometimes my children say, "I understand what I did. I just don't know how to explain it in words." How do I help them answer problems that require writing?*

"PAUSE"ATIVE BOX

Why Show Work?

Your child's teacher might request that answers be justified in writing for several reasons. Take a moment to pause and think about these reasons. I've listed some of them here. What other reasons might you think of?

- Are students thinking logically? Does the reasoning fit the situation or was a student just guessing?
- Are students using a target strategy that the teacher has identified as being appropriate for this type of problem or has a student taken a unique approach to the problem?
- Communication is at the heart of every organization—from families to boardrooms to politics. Writing down our mathematical thinking requires strong reasoning and logic but also the ability to summarize, follow a sequence of ideas, and include enough detail so that others can make sense of our sense making.

Ideas for Supporting Students' Writing in Math

TEACHING TIP

Writing down mathematical thinking is challenging for all students, but even more so for children who may be more comfortable speaking a language other than English. Here are a few tips for supporting writing in math.

- *Word Walls:* Create word walls with illustrations. A word wall is a section of a bulletin board offering a visual reminder of mathematical vocabulary for students to quickly reference. An effective math word wall contains tricky terms with a matching illustration to help students understand and connect with the terms. For example, the word *long* might be written with stretched out letters to prompt a connection to length. *Weight* might be written above a drawing of a scale. *Sum* might be indicated with a circle around the answer to an addition number sentence. *Difference* might be shown by circling the answer to a subtraction problem.
- *Sentence Frames:* Give students a lot of practice using sentence frames. Have them express their mathematical thinking out loud first. Sentence frames bring meaning to vocabulary, provide a structure for practicing and extending English language skills, and help students use the vocabulary they learn in grammatically correct, complete sentences. To get students comfortable talking and writing about triangles, for example, try these frames:

Beginning Level

This shape is not a ____________. It has ____________.

Intermediate Level

This shape is a ____________ because it has ____________ and ____________.

Advanced Level

This shape is a ____________ because it has ____________.
It is an ____________ triangle because it has ____________.

(continued)

(continued)

- *Writing Prompts:* Tailor writing prompts to students' language proficiency. In writing about triangles, for example:

Beginning Level

Build the answer into the prompt: *Write down if this is an acute or an obtuse triangle.*

Intermediate Level

Make the problem a bit more open: *How can you tell this is an acute triangle?*

Advanced Level

At this level, your prompts can be broad: *How can you tell this is a triangle?*

Using these ideas and more, your patient, supportive efforts will ensure all learners of English continue to grow as confident and competent writers.

12. Why are children asked to explain their thinking?

More important than getting the right answer, teachers want to see and hear *how* students got an answer. Maybe the student got it wrong, but in the process of explaining a solution strategy, the student can identify faulty reasoning and self-correct. That's pretty powerful meta-cognition (thinking about our own thinking). Or maybe the student got the correct answer, and in explaining their thinking, another student in the class has an aha moment. Sometimes just hearing ideas from other students clears up confusion. Explaining our thinking in math helps us build mathematical vocabulary, confidence in speaking in front of peers, and feelings of community. We're all trying to make sense of math, and our willingness to help others by sharing our ideas is part of a team effort in math class. When we work to make sense of math, we are building math-positive mindsets.

Self-Check: How Does Your Classroom Environment Support Mathematical Communication?

TEACHING TIP

If we want to build students' abilities to communicate mathematically, we must establish an environment conducive to this goal. Do you do the following in your classroom? What could you do more of?

1. *Offer Rich Tasks:* Providing rich problems that are worth discussing and have multiple entry points ensures everyone can get in on the conversation.
2. *Give Space for Constructive Struggle:* Give students time to grapple with ideas, make conjectures, and try out their thinking. Provide students with a risk-free space that gives room for a lot of divergent thinking and sometimes contradictory ideas. Develop a community of respect, tact, and good listeners. As the teacher, refrain from interrupting or supplying too much information too early. For more on constructive struggling, see Message 17, "Constructive Struggling" in *Faster Isn't Smarter* (2015) by Cathy Seeley. Seeley's resource is written for parents, teachers, and anyone interested in improving mathematics education.
3. *Use Talk Moves* (Chapin, O'Conner, and Anderson 2013): Finally, students need to be encouraged to talk out, take a chance, and engage in the discussion. Teacher talk moves like, "Who has another idea?" "What do you think of Duncan's strategy?" "Who agrees with what Ayumi said and why?" "What mathematical idea are we talking about?" and "Do you think Mateo's way will always work, and how could we find out?" can help facilitate the discourse. Most importantly, as the teacher you set the tone by modeling how to listen, respond, and respectfully question others' mathematical curiosity and reasoning. Productive noise indicates that mathematical ideas are the currency of the classroom. Give it a try! Students will grow confident in their ability to communicate their thoughts and ideas—powerful skills that will follow them throughout their math-positive lives.

13. Why does my child keep a math journal and/or portfolio at school?

A journal, or math notebook, is a record of student learning experiences over a period of time. This notebook is more than a collection of observations, facts learned, and procedures followed; it also documents student reflections, questions, predictions, and conclusions for extended problems or assessments. Children's math notebooks might also be home to student-created math dictionaries and be used as a resource for reviewing information during a math lesson.

The math portfolio is a form of authentic assessment that builds math-positive mindsets focused on progress, giving a richer picture of student's progress than a paper-and-pencil test might. The portfolio provides permanence and stability for storing and organizing students' work, reflections, and questions. This is a math vault, providing a physical copy of your child's progress in math understanding. Portfolios are often student driven and maintained, meaning your child chooses what pieces are saved in it.

"PAUSE"ATIVE BOX

Math Journal Prompts

Your child's teacher might use journals and/or portfolios as vehicles for students' self-assessment. Teachers might give five minutes at the beginning or end of math time for students to reflect in their journals, using prompts such as the following. Take a moment to pause and think about how these prompts might be helpful for your child's learning. What other questions might you add to this list?

- Which part of your homework assignment was most difficult?
- What are two questions you would like answered during today's (or tomorrow's) lesson?
- Summarize what you learned about this topic today (or yesterday).
- What is one thing that is still puzzling you?

Maintaining Math Journals and Portfolios

TEACHING TIP

Journals and portfolios are popular forms of authentic assessment in math classrooms, but how can you maintain them without losing your marbles? Here are a few tips for keeping journals and portfolios (Cutler and Monroe 1999).

- *Journals as Recording Sheets:* Have students use journals as recording sheets for learning station activities. Students leave their journals open to the pages they wrote on. They stack their open journals neatly on your desk so that you do not have to fumble through the pages. You can then go through each journal quickly during your ten-minute (sigh) lunch break.
- *Cereal Boxes:* Math portfolios can be housed in empty cereal boxes. Students bring boxes from home, cover them with paper, and decorate them with growth mindset phrases. They use the boxes to store projects, open-ended investigations, and other rich math activities.
- *Portfolio Friday:* Every two weeks, have a "Portfolio Friday" where students go through their portfolios, decide what to take home and what they want to put in their assessment portfolio.
- *Evaluation:* Evaluate the assessment portfolio as a major grade once a term. *Viola!* Authentic assessment without the authentic migraine.

14. Why is there such an emphasis on problem solving?

Problem solving is perhaps the most transformative goal of mathematics. Problem solving should permeate your child's mathematics experiences, its fibers weaving and connecting through the content being learned. What good are strong computation skills if those skills cannot be applied to real problems? We value the problem-solving process not just as a means to learning mathematics but as preparation for the challenges, setbacks, trials, and bumps of life's journey. Building problem-solving techniques and processes contributes to math-positive mindsets. Practice problem solving on a daily basis with children. One way to do this is to make activities around the house into word problems. Word problems show math is real world, relevant, and interesting. They also connect math and reading. For example, "When should we leave the house if we need to be at school by 8:00?"

"PAUSE"ATIVE BOX

Addition and Subtraction: Types of Word Problems and Ways Children Might Solve Them

Considerable research has been conducted about problem solving. Some math educators look at problem-solving strategies—from using concrete objects, manipulatives, and pictures, to counting—to represent what's happening in a word problem. Other researchers analyze the difficulty of different problem types. One of these teams of researchers has identified eleven different types of addition and subtraction word problems (Carpenter et al. 2014). The following table lists those eleven types of problems with an example for each. I've also offered a typical problem-solving approach and an equation that aligns with each problem type. Use the table to practice problem solving with children. How did it go? What ways were used to solve a problem?

Problem Type	Example Problem	How It Might Be Solved	Comments
Join			
Join: Result Unknown	McGregor has 5 baseball cards. His friend gave him 6 more cards. How many baseball cards does he have?	Make a set of 5 counters. Make a set of 6 counters. Push the sets together and count. $5 + 6 =$ ___	This is one of the simplest word problems to model (act out) with counters. Modeling the joining action is important to developing the concept of addition being joining two (or more) sets.

(continued)

(continued)

Problem Type	Example Problem	How It Might Be Solved	Comments
Join: Change Unknown	Maria has 5 marbles. How many more marbles does she need to have 13 marbles all together?	Start counting at 5 and count up to 13, raising one finger for each number (6, 7, 8, etc.) then counting the raised fingers to find the answer: 8. $5 + \_\_\_\_ = 13$	Counting from a number other than one is a skill that begins in preschool. It sure helps in this type of problem.
Join: Start Unknown	Trinh had some books. She went to the library and got 3 more books. Now she has a total of 11 books. How many books did she have to start with?	Draw 3 squares using color A. Draw 9 more squares, counting up to 12, using color B. Count the squares drawn in color B to find the answer: 9. $\_\_\_\_ + 3 = 12$	Some children might think of this as a subtraction problem. That's fine; problems that can be solved using multiple strategies build flexible thinking and math-positive mindsets.
Separate			
Separate: Result Unknown	There were 7 seals playing. Three seals swam away. How many seals were still playing?	Make a set of 7 counters. Remove 3 counters. Count the remaining counters to find the answer: 4. $7 - 3 = \_\_\_\_$	Problems about swimming away, getting lost, or being eaten are great for building ideas about subtraction, because they make sense to children.
Separate: Change Unknown	There were 18 people on the bus. Some people got off. Now there are 6 people on the bus. How many people got off the bus?	Draw 18 circles to represent the people on the bus. Cross off circles until 6 are left. Count the circles that are crossed off to find the answer: 12. $18 - \_\_\_\_ = 6$	Change unknown problems are tricky because the word *some* is ambiguous. Try forming a mental image of the action in the story to make sense of it.

(continued)

(continued)

Problem Type	Example Problem	How It Might Be Solved	Comments
Separate: Start Unknown	Miguel had some sticks of gum. He gave 2 sticks to Fritz. Now he has 7 sticks left. How many sticks of gum did Miguel have to start with?	Draw 7 rectangles to represent the gum he ended up with. Draw 2 more rectangles to represent the gum he gave to Fritz. Count all to find the answer: 9. ____ − 2 = 7	Children are sometimes stumped when *some* comes at the beginning of a problem. Working backward is a good strategy for this type of problem.
Part–Part–Whole			
Part–Part–Whole: Whole Unknown	Anujin picked 3 ripe tomatoes and 11 unripe tomatoes. Oops! How many tomatoes did he pick?	Draw 3 red circles to represent the ripe tomatoes. Draw 11 green tomatoes to represent the unripe tomatoes. Count all to find the answer: 14. 3 + 11 = ____	The idea that a *whole* (tomatoes) can have *parts* (ripe or unripe) is a big idea in math—composition.
Part–Part–Whole: Part Unknown	Knox has 9 pens. Five are red and the rest are blue. How many blue pens does Knox have?	Draw 9 rectangles. Color 5 red. Color the rest blue and count to find the answer: 4. 9 − 5 = ____ Or 5 + ____ = 9	Some children might "think addition" while others "think subtraction" during this problem. Lots of math-positive thinking possibilities!
Compare			
Compare: Difference Unknown	Maurillio has 13 stickers. Knox has 18 stickers. How many more stickers does Knox have than Maurillio?	Start at 13 and count up to 18, raising a finger for each number to find the answer: 5. 13 + ____ = 18 Or 18 − 13 = ____	This problem is rich with algebraic connections. Many children will "turn it into an addition problem," which makes good sense.

(continued)

(continued)

Problem Type	Example Problem	How It Might Be Solved	Comments
Compare: Quantity Unknown	Holland has 15 crayons. She has 10 more crayons than Hina. How many crayons does Hina have?	Make a set of 15 counters. Make a set of 10 counters. Line them up and find how many don't have a partner to get to the answer: 5. 15 – 10 = ____	The matching-up strategy works well if children talk out loud while they're solving this problem. Good problem solvers often talk to themselves!
Compare: Referent Unknown	Xander went to the toy store. He counted 12 stuffed animals that were cats. He counted 8 more cats than dogs. How many dogs did he count?	Retelling this problem helps children make sense of what's really going on in the story. Their retelling might sound like this: Xander went to the toy store. He counted 12 cat stuffed animals. He counted 8 fewer dogs than cats. That means he counted 4 dogs. 12 – 8 = ____ Or ____ + 8 = 12	The structure of these problems is kind of odd. But retelling it in your own words helps. Feel free to adjust wording to a more familiar sentence structure. The retelling just needs to mean the same thing as the original problem.

"PAUSE"ATIVE BOX

Multiplication and Division: Types of Word Problems and Ways Children Might Solve Them

Multiplication and division have their own types of word problems (Carpenter et al. 2014). I've identified them here. Look across the rows to see how multiplication and the two types of division relate to one another. Having a problem-centered classroom or home supports math-positive mindsets by providing children with real-world contexts for math. So, don't be flustered by the word problems children are working on; those problems are helping children build fluency with number, flexibility with strategies, and math-positive, problem-solving confidence.

	Multiplication	**Measurement Division (find the number of sets)**	**Partitive Division (find the number in each set)**
Grouping	Kwame has 3 packs of markers. There are 8 markers in each box. How many markers are there all together?	Kwame has some markers. There are 8 markers in each box. All together there are 24 markers. How many boxes does Kwame have?	Kwame has 3 boxes of markers. There are the same number of markers in each box. All together there are 24 markers. How many markers are in each box?
Rate	Keisha runs 4 miles an hour. How many miles does she run in 6 hours if she keeps a steady pace?	Keisha runs 4 miles an hour. How many hours will it take her to run 24 miles?	Keisha runs 24 miles. It took her 6 hours. If she runs the same speed the whole way, how far did she run in one hour?
Unit Price	Cakes cost 13 dollars each. How much do 5 cakes cost?	Cakes cost 13 dollars each. How many cakes can you buy for $65?	Jake bought 5 cakes. He spent a total of $65. If each cake cost the same amount, how much did one cake cost?
Multiplicative Comparison	The boa constrictor is 8 times as long as the garter snake. The garter snake is 2 feet long. How long is the boa constrictor?	The boa constrictor is 16 feet long. The garter snake is 2 feet long. The boa constrictor is how many times longer than the garter snake?	The boa constrictor is 16 feet long. He is 8 times as long as the garter snake. How long is the garter snake?

15. What makes a person a good problem solver?

Because it plays such an important role in mathematics, problem solving has been studied a lot by researchers. Some of this research is dedicated to comparing the attributes of poor problem solvers and successful problem solvers (Schoenfeld 1985; Silver 1985; Krulik and Rudnik 1998). Following is a bit of what we know about poor problem solvers as compared to successful problem solvers.

Poor Problem Solvers . . .

- don't plan carefully, jumping to a strategy on the basis of just a few clues;
- often don't identify relevant information, but even when they do, they don't bring the information to bear on solving problems;
- view mathematics as primarily based on memorization;
- tend to go right to looking for the numbers in the problem, pulling them out, and choosing some operation for solving them;
- focus on cue words like *fewer*, *less*, and *more* rather than the context and meaning of the problem; and
- have difficulty generalizing across problems or seeing threads that connect ideas within mathematics.

Whew! Enough about poor problem solvers. Let's focus now on what makes a child a good problem solver.

Successful Problem Solvers . . .

- have a desire to solve problems;
- are extremely persistent;
- have a variety of methods of attack;
- can skip steps when working while maintaining accuracy;

- make connections by recalling similar problems;
- differentiate between irrelevant and relevant details;
- generalize findings;
- aren't afraid to guess but don't get upset when guesses don't work out;
- hold conversations with themselves; and
- do more rereading, rechecking, and reviewing than poor problem solvers.

Good mathematical problem solving develops habits of mind that transfer to the real-world problems we all encounter.

"PAUSE"ATIVE BOX

Thinking about Problem-Solving Habits

Ideally, children translate problem-solving habits of mind to real-world dilemmas. To do so, however, children must have healthy problem-solving habits. Take a moment to pause and think about the following list. Which characteristics resonate with your child's learning? What characteristics might you want to speak with your child's teacher about?

To develop problem-solving habits of mind, children must . . .

- view the problem as important,
- accept that a problem can be approached in multiple ways using different solution strategies,
- defend their own viewpoints while respectfully listening to others' ideas,
- be willing and able to communicate about the problem verbally and/or through the use of models,
- connect the problem to other important mathematical ideas and to their own experiences, and
- see themselves as powerful problem solvers who do not give up but rather persevere when solving problems.

16. How is my child's teacher helping my child learn to be a good problem solver?

Problem solving is life! When we equip children with mental tools for approaching and solving problems, they build math-positive mindsets and are prepared to take on life's challenges. There tends to be three approaches to teaching problem solving:

- Teaching *for* problem solving.
- Teaching *about* problem solving.
- Teaching *through* problem solving.

Each approach has a role in helping your child become a good problem solver with a math-positive mindset. Let's take a deeper dive into the pros and cons of each.

Teaching *for* Problem Solving

Teaching *for* problem solving starts with learning a skill, practicing it, and then applying it to problems. For example, the teacher might show your child how to find equivalent fractions. Then your child practices that skill, and finally applies it to word problems that—surprise—can best be solved by using equivalent fractions. This approach to teaching problem solving is efficient, but it sends a message that there is one specific strategy that should be applied to certain types of problems. It decreases students' responsibility for learning and might discourage them from trying alternative approaches to problems in the future (Van de Walle, Karp, and Bay-Williams 2019).

Teaching *about* Problem Solving

The second approach to teaching problem solving, teaching *about* problem solving, acknowledges that problem solving is a process that can be complex and difficult to learn. It involves strings of tasks and thought processes that form a set of heuristics. These heuristics are suggestions and questions to follow and ask in order to resolve a dilemma (Krulik and Rudnik 1988). Several people have put forth workable problem-solving heuristics. I like to represent one widely adopted approach, developed by mathematician and mathematics educator George Polya (1957), as a cyclical process, shown here.

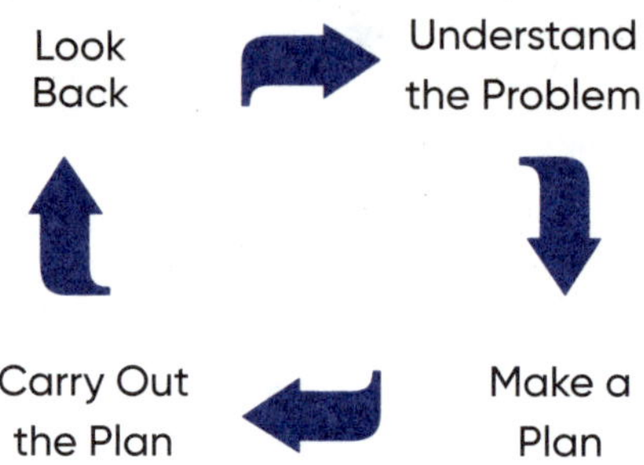

Polya's Problem-Solving Process

Teaching the phases of the problem-solving process helps students to use reasoning and to approach the problem logically. Teachers might also explicitly teach specific strategies for solving challenging problems. Some of those strategies might include drawing a picture, acting out the problem, using a model such as manipulatives, looking for a pattern, making a table or chart, trying a simpler form of the problem, using guess and check (sometimes called predicting and checking for reasonableness), and making an organized list.

Teaching *through* Problem Solving

The third approach to teaching problem solving is my favorite—teaching *through* problem solving. This way of teaching emphasizes inquiry. Instead of following the model of direct teaching inherent in teaching *for* problem solving, the teacher steps back, taking the role of facilitator while students have the responsibility to explore, try, fail, find patterns, construct, connect, reason, persevere, engage, argue, defend, and wonder. Children learn mathematics by doing mathematics and by doing mathematics they learn mathematics (Cai 2010). Teachers select rich mathematical problems that align with grade-level objectives, provide tools and support, and ask good questions to guide learning.

Creating a Visual of the Problem-Solving Process

TEACHING TIP

I recommend that you depict Polya's (1957) process, or some variation of it, visually for students, such as on a poster for display in your classroom. When you are creating your visual, consider making it circular in its formation, like the graphic shown previously and repeated here. Problem solving is cyclical in nature; we don't always understand the problem before we begin to apply a strategy to it. Sometimes it is through solving the problem that we come to understand it. Seeing the Polya problem-solving heuristic in a circular graphic helps students who get "stuck" at the *Understand the Problem* phase get unstuck and move forward. Plus, it makes the *Look Back* phase more powerful when we see that step as essential to *Understand the Problem.*

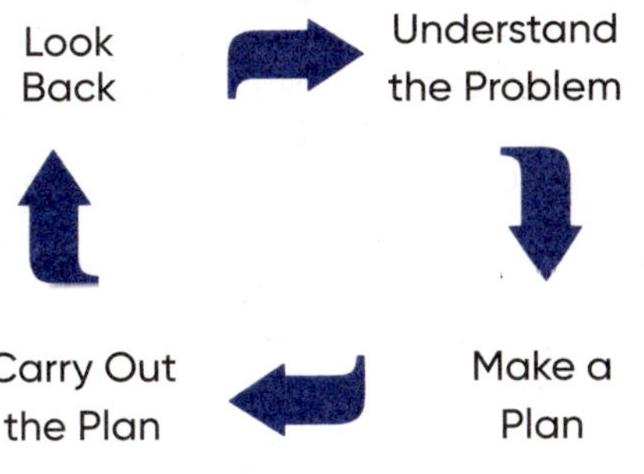

Polya's Problem-Solving Process

17. What questions can children ask themselves to help them make sense of and solve a tough math problem?

The best problem solvers have a conversation with themselves that takes them through the logical steps they follow to work toward a solution. Here are some good questions to get children talking to themselves as they work to solve math problems.

Questions to Ask at Each Phase of Polya's Problem-Solving Process

Understand the Problem

- How can I say the problem in my own words?
- What are the facts?
- What is the question?
- Are there "hidden" questions that need to be answered first?
- What additional information do I need to know?

Make a Plan

- Can I draw a picture of the situation and look for patterns?
- Can I organize the data in a list, table, or graph and look for patterns?
- Can I find other examples of this type of problem and look for patterns?
- Would it help to work backward?

Carry Out the Plan

- Have I worked on a similar problem?
- If yes, how did I solve it?
- Would using smaller numbers make the problem easier?

- Would a systematic guess-and-check approach allow me to eliminate possibilities and close in on the solution?
- Is the plan working or do I need to try something else?
- Am I getting closer to the answer?
- Do I need to separate the problem into smaller parts?

Look Back

- Did I answer the original question?
- Is my answer reasonable?
- Does my answer meet all the conditions of the problem?
- Can I solve the problem using a different approach and get the same result?

18. What is a benchmark test and how can I prepare my child to take one?

Many school districts use benchmark tests to gather data about how students are doing. They administer the tests periodically throughout the school year. Generally, benchmark tests cover the information the teacher was asked to address during a specific time frame such as six or eight weeks. Benchmark data are used to help the teacher (and the school district) know if (and, hopefully, how) students are learning. Teachers can then adjust their teaching based on that information. They can see if they need to reteach a few students, review with the whole class, or move on.

19. My child's teacher emphasizes memorizing the number facts and gives timed tests. What is the purpose of timed tests?

Timed tests are one of those unfortunate, old-school traditions that seem to stick around math classrooms despite having been shown to have negative effects on learning. Sort of like limiting recess and then wondering why kids are antsy all afternoon (come on), we sometimes do things in classrooms that don't make a lot of sense for children's development. Timed tests on basic facts have been shown to be bad for children (Boaler 2012) and actually interfere with fact retrieval and number sense (Henry and Brown 2008). Nevertheless, many teachers still rely on timed tests in math.

Teachers may have several reasons for requiring timed tests. First, we tend to imitate what has been done to us when we were students. If it was "good for us" it will be good for our children. If we grew up enduring timed tests, we think timed tests must be of value in assessing or motivating students to learn their facts. Second, the school may require the teacher to administer timed tests. Third, timed tests are easy to administer and take just a few minutes for students to complete and teachers to grade. The teacher does not have to analyze the child's strategy or reasoning, just the accuracy and number of responses. But research suggests that accuracy and efficiency in recalling the facts may actually be negatively influenced by timed testing. A study of nearly 300 first graders found that children who regularly completed timed tests demonstrated *lower* progress in knowing facts from memory than first graders who had not completed as many timed tests (Henry and Brown 2008).

Accuracy and efficiency are not the only victims in the timed testing routine. Timed tests have also been shown to elevate math anxiety among high-achieving and struggling learners alike (Ramirez et al. 2013). Students who do not know facts from memory feel like failures after a timed test and students who do know facts from memory still sometimes crack under the pressure of a ticking clock. The good news is we know that children can learn facts without the use of timed testing (Kling and Bay-Williams 2014), especially when students learn efficient strategies that they can apply to compute quickly. (For more about these strategies, see Chapter 4, Question No. 16, *What are some specific strategies, besides memorization, that can help my child compute quickly using multiplication?*)

Alternatives to Timed Tests

TEACHING TIP

Are you looking for a way to assess students' fluency with math facts but recognize that timed tests are ineffective and actually damaging to the development of fluency? Kling and Bay-Williams (2014) suggest using focused observations, interviews, journaling, and quizzes. The quizzes they recommend, however, are designed to encourage more than accuracy and are certainly not tied to speed. Instead, students show the answer and record if they "just knew it" or used a strategy. You can then review the quizzes to see which facts are automatic for each student and which strategies should be reinforced. This type of assessment offers a much richer picture of students' mastery of basic facts than a timed test. It also gives you a better idea of where to focus instruction. Plus, it encourages students to self-monitor their learning and selection of strategies, allowing them to progress toward *true* fluency with the basic facts. Using richer assessments than timed computation tests is one way to encourage the development of math-positive mindsets. Students work toward real understanding of relationships between numbers to select efficient, accurate, and flexible reason-based computation strategies.

20. Aren't kids who can do math fast considered better at math than kids who are slower at doing it?

We need a shift in our thinking. Math is not about speed! Being good at math does not require being fast at math! We need to encourage children to stick with a tough problem for a long time, working and reworking it until they find a solution. Math also isn't a one-size-fits-all, robotic exercise in which everyone spews out the same answers. Children need to grapple with math problems that have many solutions (or at least many ways of finding a correct solution) rather than relying on computation as the sole indicator of math achievement. When assessing students' mathematical abilities, we should look at fluency, which has four parts: accuracy (get the correct answer), efficiency (get it using an efficient strategy), flexibility (have a variety of strategies), and appropriateness (choose a strategy that makes sense). We will be more successful at building math-positive mindsets if we separate speed from our definitions of mathematics success. Cathy Seeley, in her books *Faster Isn't Smarter* (2015) and *Smarter Than We Think* (2014), addresses the issue of speed head-on through thought-provoking messages written for parents, teachers, and anyone interested in improving mathematics education. For more on fact fluency, see Chapter 4, Question No. 19, *I thought math wasn't supposed to be about speed. Why is it important for my child to learn basic computation facts and compute quickly?*

21. How can I prepare my child for tests that the district or state requires?

It seems like testing is never-ending. One evening at dinner, my family tallied the number of tests and quizzes our children had taken that week and the result was shocking. Why so many tests? Testing is one form of assessment. Some others are portfolios, performance tasks, observations, checklists, and anecdotal records, though these forms are less common as a means of evaluating students' progress. My hunch is that your child already gets a lot of test preparation at school. That's why you hear the phrase, "teaching to the test." Many teachers spend a good bit of time explicitly teaching test-taking strategies, typical question formats, and, of course, the objectives that will be tested. The best way to help children prepare for tests is to help them do well on daily assignments. It also helps to know what information the tests will cover. The state test is based on the state objectives. Many states publish tests from previous years on their state department of education websites. Those tests, sometimes called released tests, might help children to see the tests' formats, lengths, and questioning styles. When you talk about testing at home, keep a math-positive mindset at the forefront of the conversation. Share with children that testing is only part of their rich, broad mathematics learning experience. You aren't looking for them to be the first one done or get a perfect score. You are interested in them making progress over previous testing and moving forward in their understanding.

Is Testing Taking Over Your Classroom?

TEACHING TIP

For better or for worse, assessment often relies heavily on testing. However, keep testing in perspective. It should only be one element in your mathematics assessment plan. If you feel testing is taking over your classroom, seek out ways to replace tests with other forms of assessment. Use observation, journals, portfolios, performance tasks, and other forms of alternative or authentic assessment to evaluate students' progress in mathematics. (Read more about portfolios and journals in Question No. 13 in this chapter.) Often these alternative assessments are more student centered, rich, and revealing than paper-and-pencil tests and give you better information about how your students are progressing in mathematics.

22. My state adopted the Common Core State Standards for Mathematics (CCSSM). What do I need to know about the CCSSM?

Many people have strong feelings about the Common Core, but what is it really? The Common Core State Standards for Mathematics (CCSSM) has two main sections. First, there are the Content Standards, which are the mathematics objectives, by grade, that students should learn. The Content Standards help define the big ideas in math for each grade. (See Chapters 4–7 for more about the big ideas in number and operations,

"PAUSE"ATIVE BOX

Standards for Mathematical Practice

The eight Standards for Mathematical Practice help students become proficient mathematical thinkers and doers. Following is a summary and explanation of the Standards for Mathematical Practice. Take a moment to pause and familiarize yourself with each practice. What questions might you have for your child's teacher based on these practices?

Explanation of Standards for Mathematical Practice from the Common Core State Standards for Mathematics

Mathematical Practice	What Students Do
1. Make sense of problems and persevere in solving them.	Grapple with tough tasks to learn mathematical reasoning and problem solving. They keep working even when a problem is challenging.
2. Reason abstractly and quantitatively.	Represent problems using manipulatives, drawings, graphs, words, and symbols (including numbers).
3. Construct viable arguments and critique the reasoning of others.	Use words to discuss, justify, explain, disagree, and support their ideas. They also try to understand and question others' ideas.
4. Model with mathematics.	Apply mathematics to other subjects in school, to their previous experiences in math class, and to the real world. They can show their ideas in many ways.
5. Use appropriate tools strategically.	Have good reasons for choosing to use manipulatives, rulers, calculators, and other tools to help make sense of and solve problems.
6. Attend to precision.	Try to be clear and accurate while discussing and solving problems.
7. Look for and make use of structure.	Find patterns in problems, then apply them to new types of problems and situations.
8. Look for and express regularity in repeated reasoning.	Identify repeated calculations, then come up with shortcuts that make problem solving more efficient.

From National Governors Association Center for Best Practices and Council of Chief State School Officers (2010).

geometry, measurement and data, and algebra.) The Content Standards also ensure that the major objectives in mathematics are covered in an orderly way, with ideas building upon one another and without gaps or omissions.

The second part, the Standards for Mathematical Practice, emphasizes the thinking processes that will help students understand math deeply instead of relying on rote procedures or memorization. As a child, you may have learned math by memorization but never really made sense of it. Today's math educators know we can do better. Students now learn mathematics with understanding. What if children could decode words and pronounce them correctly when reading but didn't comprehend the meaning of the words? Comprehension involves a much higher level of thinking than simply speaking the words. Learning mathematics with understanding is essentially the same thing. No matter how well children can perform rote procedures in math, this ability is not very useful if they do not know how to apply these skills. The Standards for Mathematical Practice seek to help students develop a lifelong understanding of mathematics that is useful at home, in elementary and secondary school, in college, and in the workplace.

23. My state didn't adopt the Common Core State Standards, so how can I find out what math objectives my child is required to know?

Each state publishes their state-adopted curriculum standards on their respective public education websites. Parents can review the standards to see where math concepts fit in the sequence of instruction across the grades. Some state websites also provide plain-language explanations for the standards, opportunities for parents to comment on revisions, and ideas and materials to help parents understand and support learning at home. The Texas Education Agency website, for example, details all the proposed changes, revisions, public comments, and expert reviewers' opinions on the Texas Essential Knowledge and Skills for mathematics. It's neat to see the process as it unfolds over several years prior to the adoption of new objectives. Private schools may or may not utilize their state's curriculum standards, so asking for a printed copy of the school's curriculum can clarify the sequence of mathematics learning for children enrolled in nonpublic schools.

QUICK REFERENCE CHART

Common Concerns about Math Teaching and Learning

This table is a quick reference for some of the *most common concerns* I hear from parents regarding their children's math teaching and learning. The table lists the concern, an explanation of the thinking behind it, and an idea that addresses the concern. Consider photocopying this table and placing it somewhere you can be reminded of what is needed to build your math-positive mindset and that of your children and/or students. Also use it to help get to know your child's teacher and math classroom.

Concern	Explanation	What to Do	Question Number(s) (to Learn More)
Communicating with the teacher	Research shows that open lines of communication between home and school are beneficial to children's learning.	Your child's teacher and you can be a powerful team. Build a math-positive partnership by sending a quick appreciation email or a heads-up when things are tough. And if you see great things happening, let the principal know. That will make everyone feel good!	1, 2
Parent-teacher conferences	Conferences give teachers a chance to share students' progress and parents a chance to ask questions.	Begin the conference with a compliment or by showing gratitude to the teacher. Plan a few general questions to ask. Wrap up with an offer to help your child succeed in math. Children are the real winners when caring adults are on the same team.	3
Changes in math	The world is changing, technology demands math, old approaches weren't working, and math phobia/anxiety is barrier to success.	Explore growth mindset with this child-friendly video. Search YouTube for animated episodes of the "Class Dojo's Growth Mindset" video series.	4, 5, 6

(continued)

(continued)

Concern	Explanation	What to Do	Question Number(s) (to Learn More)
Learning with understanding	Memorizing steps or procedures results in less robust, more fragile learning that is not easily transferred to new contexts. In other words, you learn it once, but it doesn't stick.	Play games that build reasoning, involve finding patterns, and require strategizing. Try this: Roll a die three times, adding the value shown on the die or ten times its value. Try to get as close as possible to 60 without going over. Keep track in your head or use paper and pencil.	7
Mathematical processes	The processes of problem solving, reasoning and proof, communication, connections, and representation are the thinking threads that tether math to our everyday lives.	You use math! So talk with your child about it. Have you solved a real-world math problem at work or at home—ordering office supplies, measuring for a new fridge, mixing chemicals for the pool? Be math aware and math positive and your child's buy-in to the mathematical processes will grow.	8
Mathematical vocabulary	Mathematics is more than "naked numbers" or computation items on a worksheet. Math classrooms are filled with vibrant discussions, word problems, and mathematical language.	When children have a strong mathematical vocabulary, they confidently share their mathematical ideas, explain their thinking, and justify their responses. Talk about math with your child every day to build vocabulary and exercise the math-chat muscle. It will get stronger with practice.	9, 10
Showing work	Real math problems (not just recall of facts or computation) require deep thinking. Writing down mathematical thinking helps the teacher make sense of students' reasoning.	If the process of putting words on paper is a stumbling block, have children tell you about their reasoning while you jot down their ideas. Then have them write their thinking on the homework.	11, 12, 13

(continued)

(continued)

Concern	Explanation	What to Do	Question Number(s) (to Learn More)
Importance of problem solving	The meat of mathematics is problem solving. We cannot use mathematics meaningfully without a worthy cause—engaging, messy, rich, real-world problems.	There are a lot of online resources for math riddles that require problem solving. Try www.riddles.com or www.wodb.ca for puzzling challenges your child and you can solve together.	14, 15, 16, 17
Word problems	Show math is real-world, relevant, and interesting. Connects math to reading.	Make daily activities into word problems. "How many tortillas will we need for the breakfast tacos? How many eggs? And don't forget the bacon!" For a chart of different types of word problems, see Question No. 14.	14
There is more than one way	There are multiple ways to arrive at an answer when solving math problems.	How many ways can your child approach a problem? Ask, "What's another way you can think of to solve this problem?"	15
Memorizing versus knowing basic facts	To memorize a fact, a child may or may not understand it. To know and make sense of a fact is to understand how it relates in a larger scheme.	Make a personalized set of basic fact flashcards. Toss out the ones your child "knows" and focus on understanding the rest.	19, 20
Speed and timed tests	Timed tests are not supported by research as a way to learn or assess basic facts. Math anxiety causes interference with memory, decreasing speed and accuracy and causing math phobia.	It helps to know some basic facts from memory. However, rather than emphasizing speed, work with your child to build strategies like Number Bonds for the Number Ten, Use a Known Fact, Number Splitting or others (see Chapter 4 for an explanation of each of these strategies). The goal is to think flexibly about numbers so we can use them in all sorts of ways.	19, 20

(continued)

(continued)

Concern	Explanation	What to Do	Question Number(s) (to Learn More)
Preparing for tests	Testing is one element of a full assessment plan. Along with assignments, projects, interviews, and portfolios, tests provide a snapshot of mathematical understanding.	Keep conversations about testing light and positive. Help your child stay on track by doing well on daily assignments. Know what information the test will cover and review as needed. Encourage progress over previous testing rather than a perfect score or being the first one done.	18, 21
Fact fluency	Fluency has four parts: accuracy (get the correct answer), efficiency (get it using an efficient strategy), flexibility (have a variety of strategies), and appropriateness (choose the best strategy).	For fact fluency practice, play *The Sum What Dice Game* (Stenmark, Thompson, and Cossey 1986). Each player writes out digits 1–9 on a paper. Take turns rolling two dice, covering either the sum or any two uncovered numbers that add to the sum. When a player cannot play, he or she adds the total of all remaining uncovered digits. The rest of the players continue. The player with the lowest score wins.	20
Common Core and state standards	Standards are intended to ensure that students learn mathematics in an orderly and logical sequence without skipping over important ideas they'll need later on.	Each state chooses which standards they adopt. You can find out what math standards are being used in your state by searching online. You might want to print them out and look them over. It's eye-opening to see how mathematical ideas build on one another.	22, 23

Apply Your Math-Positive Mindset

Now that you're on your way to developing a math-positive mindset and you're working closely with your math-positive team (parents and teachers alike), it's time to support children's math learning by encouraging a positive math environment at home and taking a deeper dive into the math content being learned.

(continued)

Chapters 4 through 7 offer specific advice for the content areas in math—Number and Operations, Geometry, Measurement and Data, and Algebra. These chapters break down confusing concepts, tackle common misconceptions, and present ways parents and teachers can reinforce mathematical vocabulary while supporting conceptual learning over memorized procedures. They are filled with simple explanations for some of the tricky, modern incarnations of the math you learned in elementary school and plenty of math-positive, attitude-building activities.

I Can Encourage Math Learning at Home

CHAPTER 3

(continued)

(continued)

(continued)

(continued)

1. What are the purposes of math homework?

Not all parents support daily homework. I admit I've cursed more than a few homework packets. Not all students believe there is a benefit to completing homework assignments. I confess my children have put up a fit (or two or twenty!) over the years. Not all teachers assign daily homework. I declare a dislike for grading homework. Nevertheless, math teachers almost universally assign homework and do so for several reasons:

- First, homework reinforces skills and concepts in math—sometimes new but usually for review.
- Second, students may not have time to complete work at school, necessitating completion at home.
- Third, teachers may view homework as a means of helping parents stay in touch with what is going on in math.
- Fourth, teachers may assign homework in an attempt to teach children responsibility.
- Finally, teachers may view regular homework as a means of developing good study habits in children.

TEACHING TIP

Communicating the Value of Homework

Be open with parents about your expectations for how long homework should take to complete and why you are assigning it. Explain the value of homework as a tool for building math understanding but also for building responsibility and good work habits. When parents understand your expectations and why you assign homework, they are more likely to buy-in to the idea. Support ongoing, open communication with parents through regular emails, newsletters, and family math nights.

2. How long should I expect math homework to take?

The right amount of math homework can foster positive character traits such as independence and responsibility, time management, and the understanding that math takes place everywhere. But too much math homework can also have negative effects on children. Children may grow bored or resentful if they are required to spend too much time on schoolwork. Homework may prevent children from taking part in healthy or special activities and lessons that also teach important life skills. Children who are overloaded and overwhelmed by homework may resort to cheating, either by copying classmates' assignments or other means.

The younger the child, the less time the child should be expected to devote to homework. A general rule of thumb is that children do ten minutes of homework for each grade level. So, first graders should be expected to do about ten minutes of homework, second graders twenty minutes, third graders thirty minutes, and so on. If children are spending more than ten minutes per grade level on work at night, then you may want to talk with their teachers about adjusting the workload (Dawson n.d.).

3. Is it best to do homework immediately after my child gets home from school?

Children's physical and emotional signals should tell you the optimal time for homework. In our family, we tackle homework right after school. This frees up time later for relaxation, music or dance lessons, sports, and other activities. However, children may have difficulty focusing on homework after a full day of school, so give them time to do something relaxing, have a snack, and tell you about their day at school before starting homework.

Thinking Critically about Homework Assignments

TEACHING TIP

Look at the homework you assign with a critical eye. Is it fulfilling your intended purposes? Ask yourself questions like the following:

- How does the homework assignment support the learning happening in the classroom?
- How does the homework assignment develop problem solving, reasoning, and mathematical communication?
- How does the homework assignment build positive attitudes about mathematics?
- How does the homework assignment involve families in appropriate ways? How is it meaningful to children and families?
- How could the homework assignment be improved to do these things better?

When we assign homework, we should consider ways to build (1) conceptual understanding of mathematics, (2) confidence in mathematics, (3) positive attitudes about mathematics, and (4) strong school-home connections.

4. How can I motivate my child to complete homework before doing other activities?

Lack of academic motivation is pretty normal among children. If children are not excited to do math homework, they are not bad kids and you are not a bad parent. You may have to be a bit creative to jump-start children's intrinsic motivation.

Parents often want to put a carrot in front of their children to move the homework along. Experts are divided when it comes to using rewards such as money or extra screen time for completing homework. Many prefer that parents offer a hug or encouraging comments such as, "I bet you're looking forward to having free time to play when you finish your homework. It's good to finish quickly so you have more time with your friends."

I confess. When my son Duncan was in first grade, I occasionally motivated him to work efficiently by putting a small cup of chocolate chips beside him. When he completed a row on his worksheet, he could take a chocolate chip. Slices of apple or whole wheat crackers would certainly have been healthier, and some experts say food should never be used as a reward as it may lead to later health issues. If you do choose to use some sort of extrinsic motivation for completing homework efficiently and without complaint, consider the reward of time—offer to take a walk after dinner together, spend a few minutes playing a favorite board game together, or allow children to choose some other activity that they enjoy.

5. What if my child's homework seems too easy or too hard?

For simplicity, most teachers assign the same homework for all students in their class. But if you feel that your child's homework is too difficult (tears, grumbles, high-anxiety stuff) or too easy (done in ninety seconds), you might want to put a bug in the teacher's ear. It's up to teachers, though, to assign the homework. So, if the teacher doesn't want to adjust it, you're kind of stuck. If the homework is too difficult, try using some of the stress-relieving tips outlined under Chapter 1, Question No. 7, *How do I ease my child's math anxiety?* You might also try the "brain breaks" listed in this chapter under Question No. 11, *How can I make homework time playful and stress free?* If the homework is too simple and you want to add some challenge, try some of the online programs and apps listed in this chapter under Question No. 25, *What online math assistance is available for my child and for me?*

TEACHING TIP

Communicating with Parents about Homework

If parents question you about homework assignments, don't feel defensive or second-guessed. Be glad you have parents who are engaged and interested in their child's learning. Generally, a quick note or email can clear up confusion. If a face-to-face meeting is necessary, keep the conversation light and positive. Tell parents you are glad that they are interested in mathematics. Listen. Offer empathy and support. Look at it as an opportunity to share how excited you are about math. When explaining the assignment, speak to the parent as a peer. Be respectful and patient. This is a chance to gain a friend and an ally in the quest for higher achievement in mathematics and stronger home-school connections.

6. How should I set up a homework space?

Left to themselves, many children would do homework in front of a TV under no supervision. At school, children generally don't do classwork on the playground or in the lunch room. Similarly, you need the right environment for home learning. A dedicated homework spot limits kids' distractions, helps them focus on completing work efficiently, and tells them that you believe homework is important. If children complete their homework at an after-school program, it is still good to have an area at home where any additional work can be done.

In our family, our younger children do homework in a spot where it is easy for me to be nearby and provide help. They usually sit at a corner of the kitchen table while I wash never-ending dishes or prepare dinner. As our children have gotten older, I've placed a desk in their bedrooms to encourage independence. I try to make sure I peek in periodically to help them stay focused and to offer encouragement. Be sure to keep math tools such as a calculator and sharp pencils handy so that children do not have to spend time searching for them.

7. I know my child uses a variety of math tools at school. Should I have tools available for homework, and if so, what tools?

Make math activities, well . . . active. Math should not be a passive activity. Children learn best when they can touch, move, and "build" understanding of math concepts (Freer Weiss 2006). Educators often make math physical through tools called manipulatives. Manipulatives are physical objects that children can use to construct their own mental images of the abstract mathematical ideas and processes they are learning (Van de Walle, Karp, and Bay-Williams 2019). Besides aiding in understanding, manipulatives get children excited about math. Students who use manipulatives report that they are more interested in mathematics, and research shows that long-term interest in mathematics translates to increased mathematical ability (Sutton and Krueger 2002).

You may choose to purchase math manipulatives from a school supply store, catalog, or website for children to use during homework time. Math manipulatives are a regular holiday and birthday gift in our home and are kept on the toy shelf to encourage frequent use in play and learning.

"PAUSE"ATIVE BOX

Cool Math Tools (That Might Make Great Birthday Gifts!) for Children

Following are some of my favorite math manipulatives to have on hand at home. Take a moment to pause and check out how to use these tools. Which ones are familiar to you and/or might you have at home? Which ones are new to you? Show your child the following—which ones do they use in school?

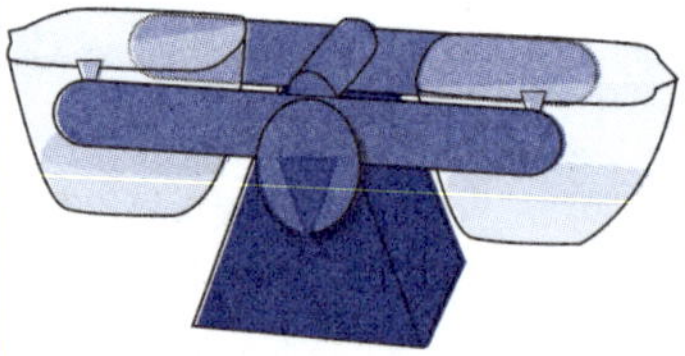 Pan Balance	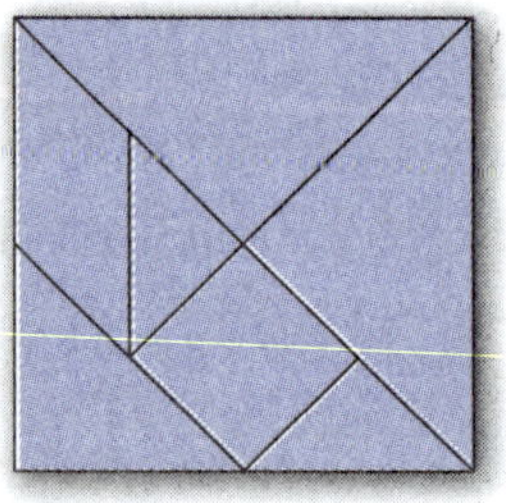 Tangrams	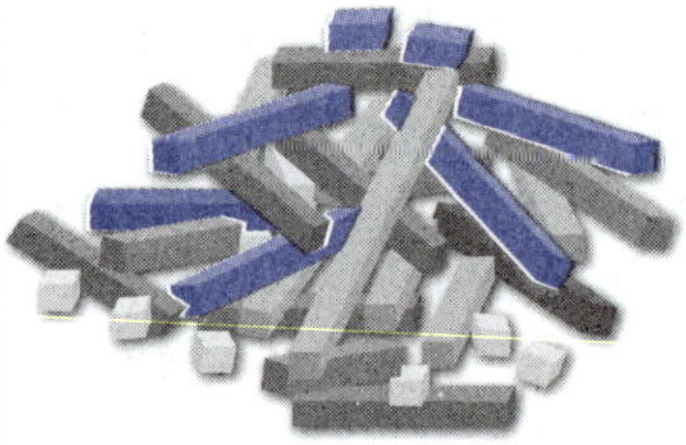 Cuisenaire Rods
 Pattern Blocks	 Base Ten Blocks	 Color Tiles
 Inch Cubes	 Multisided Dice	 Algebra Tiles

(continued)

(continued)

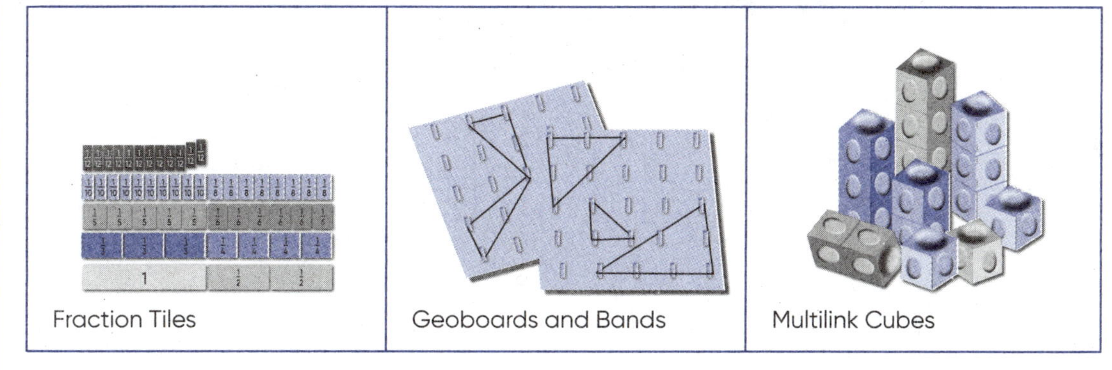

Fraction Tiles | Geoboards and Bands | Multilink Cubes

8. What if I don't have money for fancy-schmancy math manipulatives? What are some activities I can do with everyday objects?

In addition to commercially produced manipulatives, with a little creativity, everyday household items can become powerful mathematics tools. Simple objects such as buttons, cereal, dry macaroni, and toys such as cars, blocks, and stuffed animals encourage children to explore new concepts, make valuable connections, and build strong foundations in math. Whether they are made specifically for math (like geoboards) or for other purposes (like beans), using manipulatives helps children to "build links between the object, the symbol, and the mathematical idea both represent" (National Research Council 2001, 354).

"PAUSE"ATIVE BOX

Math with Everyday Objects

Math manipulatives are a lot of fun and can give children hands-on practice with mathematical concepts, but you don't have to spend a lot of money to learn about math. Here are two simple activities to get you started. Take a moment to pause and play one of these games with your child.

Game 1: Button Fractions

This game, ideal for children in grades kindergarten through third, teaches attributes, classification, and fractions using a set model. To play, you simply need an assortment of buttons (different sizes, colors, shapes, thicknesses, number of holes, etc.).

Instructions

1. Allow your child to explore the assorted buttons.
2. Listen for your child's comments about the buttons' attributes. For example, your child might point out the number of holes a button has, its color or thickness, or if it's made of metal, wood, or plastic.
3. Gather the buttons in a pile and select four buttons that share at least one attribute. Ask, "What do these buttons have in common?"
4. Choose a new set of four or five buttons and look for similarities.
5. Play a "one of these things is not like the others" game with the buttons. Make a set of four or five buttons—one of which is dissimilar in some way. Ask, "Which button does not belong and why?"
6. Build mathematical vocabulary by saying, "Four out of five buttons are made of plastic. One out of five buttons is made of metal."
7. Have your child create a "one of these things is not like the others" set and try to stump you.
8. Encourage your child to use math language to describe the fractional makeup of the sets of buttons.
9. If desired, your child can record descriptions of the sets of buttons on paper by drawing a picture of the set of buttons or writing, for example, "Four out of five buttons are round. One out of five buttons is a square."

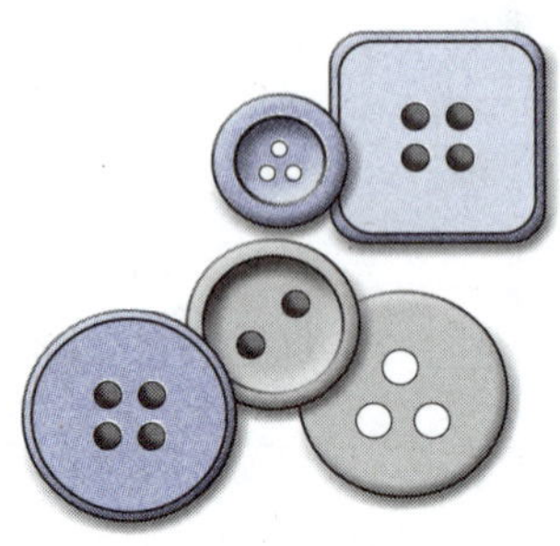

(continued)

(continued)

Game 2: Coin Trading

This game from Marilyn Burns's book *About Teaching Mathematics* (2015) teaches coin values, counting, and addition. To play, you'll need one die, paper, and a pile of pennies, nickels, and dimes.

Instructions

1. Each player draws a three-column grid on their paper and labels the columns *Penny*, *Nickel*, and *Dime*. This is their game board.
2. Put the pile of coins in the center of the table.
3. Take turns rolling the die and adding the corresponding number of pennies to your board. For instance, if Player A rolls a 4, the player takes four pennies and puts them on their board in the penny column.
4. The idea of the game is to show, in any column, the smallest number of coins possible. So as soon as you have enough pennies to "trade up," you may do so. For example, if you have five pennies, trade them for a nickel and place the nickel in the nickel column. As soon as you have two nickels, "trade up" for a dime and place the dime in the dime column, and so on.
5. The first player to have five dimes in the dime column is the winner.

Here is a sample game where the player is just 12 cents away from victory!

9. What are virtual manipulatives and how can they help my child?

Virtual manipulatives are essentially digital versions of the physical manipulatives we use when learning math. Using a touch screen device, children can move, flip, spin, shuffle, group, count, and otherwise manipulate the virtual objects (Burris 2013). Common virtual manipulatives include base ten blocks, geoboards, tangrams, coins and pattern blocks. Many are available online at no cost.

Virtual manipulatives have been shown to share many of the same positive effects on children's learning as physical manipulatives (Moyer-Packenham and Westenskow 2013). But they are also unique. Following are some of the benefits of using virtual manipulatives.

- First, using virtual manipulatives allows children to follow their mathematical thinking processes on the screen, supporting their reasoning while allowing them to retrace steps in their thinking, thus limiting cognitive overload (Suh and Moyer 2008).
- Second, using virtual manipulatives saves time over drawing a sketch or model, allowing children to focus on the math connections and relationships rather than whether or not their giraffe and hippo are drawn to scale or look more like a llama and a rhino (Moyer-Packenham, Salkind, and Bolyard 2008; Suh and Moyer 2008).
- Third, some online programs that use virtual manipulatives can be set to respond to an individual child's learning pace.
- Fourth, virtual manipulatives are a portable, clutter-free, low-cost alternative to physical manipulatives. You can download apps for a nominal fee or, in some cases, free of charge. Children can then use these apps while riding in the car or sitting in the orthodontist's chair.
- Finally, let's face it. Kids these days seem to be hard-wired for technology. They can't get enough of the stuff! Virtual manipulatives may engage your child in mathematics in ways that plastic cubes and paper number lines do not.

TEACHING TIP

Sharing Virtual Manipulatives with Parents

Keep parents up-to-date on the latest virtual manipulatives available for learning mathematics so that they can use the tools at home. Several organizations are reputable sources, including the National Council of Teachers of Mathematics. If your school-adopted textbook includes online support, make sure parents know how to access those materials, especially if the resources are password protected.

"PAUSE"ATIVE BOX

Virtual Manipulatives Your Child Can Show You How to Use

Take a moment to pause and, with your child, explore the following options for virtual manipulatives online. What options might you want to add to this list?

1. *MATTI Math.* This enhanced offline version of the award-winning National Library of Virtual Manipulatives is available as a downloadable app for a nominal fee. Available in English, Spanish, Chinese, and French. Want to give simple coding a shot? Try *Ladybug Leaf*, found in the Measurement section. Don't have a geoboard handy? No problem. The virtual Geoboards applet, located in the Geometry section, offers explorations with shapes, perimeter, and area.
2. *National Council of Teachers of Mathematics' Illuminations.* The ad-free *Illuminations* site, sponsored by the National Council of Teachers of Mathematics, offers many virtual manipulatives; some require membership in NCTM (which is open to anyone). The manipulatives are searchable by grade and content. Click on the Interactives button, then choose the appropriate grade band (PreK–2, 3–5, etc.). Try *Coin Box* to practice adding and subtracting coins in a fun, easy-to-use game. Geometric Solids offers mind-blowing visuals of 3D shapes and the 2D nets that fold to make them. Try using the shading option to help you keep track while counting the edges, faces, and vertices. I also love using Pan Balance-Numbers to explore the meaning of the equal sign by using a pan balance. It illustrates how the numbers (or equations) on one side must balance (or be equal to) the other side.
3. *Houghton Mifflin Harcourt.* This learning company offers many virtual manipulatives, including number lines that can be used for addition, subtraction, and multiplication. Also try the base ten blocks to explore number composition with place value, number comparisons, and operations.

10. What about using calculators? Isn't using a calculator cheating?

Parents often express confusion about the role of calculators in the elementary classroom. The topic can be divisive and emotion-laden. But I don't think it has to be. When I teach my university elementary education majors about calculators, we do a little experiment (Cutler 2005). It goes something like this: I begin by asking my students about their preferred computation strategies—which approaches they find most efficient and effective. We list the strategies on the board and group them into four general categories: manipulatives (concrete), paper and pencil, mental math, and calculator. Next, I divide the stu dents into groups of five and assign each person a job: one person uses the calculator, another uses mental math, another uses paper and pencil, and another uses base ten blocks to represent and solve computation items. The fifth member of the group serves as the judge and recorder, responsible for identifying who solved the problem most quickly and keeping a tally of the results for each problem.

Next, I display one arithmetic problem at a time (see the problems that follow) and remind participants to solve the problem quickly using *only* their assigned strategy. That means if your assigned strategy is to use the calculator, you must punch in each digit; if your strategy is paper and pencil, you must write down the entire problem, and so on. Look over the arithmetic items. Can you guess which ones are solved most quickly with each strategy?

a. $23 \times 10 =$
b. $587 + 161 =$
c. $15 \times 6 =$
d. $490 \div 7 =$
e. $325 - 249 =$
f. $100 \times 89 =$

Following the experiment (and a flood of giggles and groans from the groups), we discuss our results. My students frequently comment on their surprise in seeing that mental math is not the quickest approach for every arithmetic problem. Often, those who must punch a memorized known math fact into the calculator or lay out an absurd number of manipulatives express frustration with the unreasonableness of their assigned method. We generally conclude the experiment with a consensus that children need a variety of strategies from which to wisely select and plenty of opportunities to hone their decision-making skills.

If the goal of a math activity is to teach children mental math strategies or paper-and-pencil procedures, then using a calculator would not make sense. But if the goal of an activity is to help children solve a rich, real-world problem with authentic numbers and multiple calculations, using the calculator enhances the overall mathematical learning.

Rich, messy, real-world problems that genuinely intrigue students are the best fodder for learning math. For example, third graders might be excited to help plan the school's family math night. They know they want to offer families pizza and soda, but how much should they order? What wonderful math questions arise as they consider how many people might

"PAUSE"ATIVE BOX

Thoughts about Calculators in Elementary Schools

Following are some thoughts about the use of calculators. Take a moment to pause and reflect on each statement. Do you agree with the statement? Why or why not? What statements might you add to this list?

- Calculators used as tools are powerful.
- Calculators used as crutches are unproductive.
- For rich, real-world problems, the calculator might allow for the use of authentic data that would otherwise be beyond children's reach.
- Teachers must incorporate open-ended math explorations (not just basic computation) for children to use calculators in the most valuable and appropriate ways.
- With practice, children can make judgments about when it makes good sense to use a calculator.

attend, how many slices of pizza they will eat, how many slices of pizza in a whole pizza, how many ounces of soda they will guzzle, how many ounces of soda in a bottle, the price of bottles versus cans, and so on. As my son Duncan reminds me, "Use problems that fit in our world. We like to learn about stuff we care about. Movies and television shows and pizza—that's what we like. Maybe sports, too, but mostly pizza."

Where the numbers and computations involved in this meaningful, real-world problem might be out of reach to third graders using manipulatives or paper-and-pencil computation, the calculator can help students focus on what calculations to perform, not just how to perform those calculations, thus enhancing exploration, problem solving, and real-world applications (Reys and Arbaugh 2001).

11. How can I make homework time playful and stress free?

Don't let homework become a shouting match or power struggle. (Cringe!) Keeping homework sessions lighthearted may help limit frustration for both your child and you. Rather than a time of drudgery, keep math positive and interesting. In my son Duncan's words, "We like it when math is made to be fun." For instance, when Duncan brought home a worksheet of subtraction items, we "warmed up" by playing a quick game with dice. Each of us rolled two dice and Duncan found the difference of each set of dice by subtracting. Then Duncan determined which of us had the greater difference and gave that player a point. The first player to earn seven (Duncan's lucky number) points was the winner. The winner got a chocolate chip, of course.

Another way to keep it playful and stress-free is by taking a short break as needed. When you notice children's attention waning, offer a ten-minute intermission. Choose something to do together—make a snack, talk about fun plans for the weekend, or get the blood pumping and endorphins flowing by dancing to a silly song. After a few minutes, tackle the next section of homework. In time, you may find that children can tolerate longer periods of concentration before needing a break.

12. My children get so frustrated that they sometimes cry during homework. How much frustration is normal? Am I creating negative associations by making them work through tears?

My own children have been known to shed a few tears of frustration during challenging homework. But if children are crying daily during homework time, you must address the situation. First, consider children's overall emotional and physical health. Are they getting adequate sleep? Are they overscheduled with activities and lessons? Could they be suffering from anxiety, depression, or some other serious mental health challenge? Second, could you be inadvertently putting pressure on your child by overemphasizing grades over effort? Third, is the homework environment or schedule out of line with your child's needs? Are there too many distractions during homework or does your child need more downtime first?

Once you've addressed all of the concerns, try these approaches: Start with the easiest problems first to build children's confidence. Break homework sessions into manageable time periods—work for ten to fifteen minutes, then take a five-minute brain break. Use encouraging comments such as "That was a difficult problem, but you kept working on it until you found a solution. I appreciate your effort!" (See Chapter 1, Question No. 8, about praising effort.) Remember to keep homework time light and as stress free as possible.

"PAUSE"ATIVE BOX

Ideas for Brain Breaks

Following are ideas I've found helpful in giving my children "brain breaks." Take a moment to pause and think about these ideas. Which ones would you like to try? What additional ideas might you have?

- *Bubbles:* Children of all ages love to blow and pop bubbles. Have children blow gently, then count aloud to see how long they can blow before the bubble pops or lifts off from the wand. Count how many tiny bubbles they can make with one wave of the wand. Count how many bubbles they can pop in ten seconds.
- *Stuffed Animal Balancing:* Have children put a small stuffed animal on their head and then let it drop into their hands. Count how many steps children can take while balancing the stuffed animal on their shoulder. For an added challenge, have children walk backwards or slide sideways.
- *Cotton Ball Hockey:* Use a drinking straw to try and blow a cotton ball to the end of the table. Count to see how many breaths it takes. Use several cotton balls and see who—your child or you—can get the most cotton balls into the other person's end zone.
- *Measured Movements:* Marching in place, have children show how fast their feet can move, then have them show how slowly their feet can move. Use measurement words to describe the marching—heavy, light, tall, and short. March to the beat of a song, then march twice as fast.
- *Name That Tune:* Hum a few measures of a familiar children's song. Have children guess the song. Then sing and dance to the whole song together.

13. How can I help my child to persevere when math is challenging?

Struggling at times is normal and is even necessary to, and valuable in, understanding mathematics (Mirra 2004). The National Council of Teachers of Mathematics (2014) has adopted the phrase *productive struggle* to signify a challenging task that requires students to persevere in the face of difficulty. Productive struggle is tied closely to growth mindset (see Chapter 1). When children encounter a challenge that requires extended effort, it might help to compare it to other endeavors that require hard work and persistence, such as playing a sport or a musical instrument. How many practice shots did your child's favorite basketball player take before winning the slam dunk contest? (Hey, that's a math question.) Athletes do not become great without effort and struggle. Likewise, you can bet Yo-Yo Ma's first attempts at playing the cello were a bit screechy. Perseverance doesn't indicate weakness or a lack of ability. Perseverance is what separates experts from novices. To struggle is to learn!

Acknowledge the fact that mathematics can be challenging at times. Remind children that persistence and hard work are the keys to success. When children are struggling with a difficult concept, use encouragement to support their efforts; don't just praise their innate ability. Experts in the field of children's self-esteem say that praising children's effort ("You worked hard on that problem.") encourages them to adopt incremental motivational frameworks (for additional suggestions on what to say to praise effort, see Chapter 1, Question No. 8, *I want to be supportive of my child, but I am not a fan of praise since it seems to turn my child into a bit of an "approval junkie." How can I build my child's confidence in appropriate ways?*). Over time, children begin to believe that ability is malleable, success is a result of hard work, and challenges can be enjoyable. In contrast, praising children's inherent abilities ("You're so smart.") encourages them to believe they are either born with the ability to do math or not born with the ability to do math (Gunderson et al. 2013). Reinforce feelings of mathematical power versus learned helplessness. To foster continual effort and growth, the National Association for the Education of Young Children recommends that parents notice and give feedback about determination rather than results (Young and Reed 2017). When parents promote perseverance, children feel safe to try and fail and try again.

One afternoon, my son Duncan agonized over a particularly challenging word problem. I sat beside him, encouraging him to use a variety of strategies. He made a list. He drew a picture. He used mental math. Finally, he threw up his hands and yelled, "I can't be a problem solver today, Mom. This problem is unsolvable." I resisted the urge to rush in and do the work for him. Instead, I gave him a break and a snack. Ten minutes later, Duncan called me over to show me the completed problem. I said, "See, it wasn't unsolvable after all. You just had to use every strategy you could think of. All that work paid off." When children feel supported, they are empowered to persevere.

14. Sometimes my children say, "I understand what I did. I just don't know how to explain it in words." How do I help them answer problems that require writing?

Math isn't just numbers. One of math education's major goals is to help children communicate their reasoning as they justify and explain strategies and ideas. This communication may be oral or written (NCTM 2000). Teachers incorporate writing in math class to help students deepen their understanding of important concepts and connect math to their own experiences and the real world. When children write, they must revisit their thinking and reflect on their ideas. Student writing gives teachers a way to assess if and how their students are learning (Burns 2004).

Communicating mathematically involves numbers, words, and sometimes pictures. Writing about math flows naturally from talking about math, so have children discuss their ideas before they begin to write. Using sentence frames can help children get started. (For more insights on writing in math, see Chapter 2, Question No. 11, *Why are children asked to show their work? My children can do it quickly and accurately in their heads. It seems like all that writing is just slowing them down.*)

"PAUSE"ATIVE BOX

Sentence Frames for Sharing Math Thinking

Take a moment to pause and think about how the following sentence frames might help children explain their mathematical thinking. What other sentence frames might be helpful?

- I needed to find out ________________. I figured it out by ________________.
- I wanted to know ________________. I started by ________________. Then I ________________.
- I think that the answer is ________________. My reasoning is ________________.
- To start the problem, I ________________. My next step was ________________.
- This is a picture that helped me make sense of the problem: ________________. I used the picture to show that ________________.

15. I think my child understands the math concepts, but I'm not sure. What questions can I ask to check for understanding?

Questioning is a good way to determine to what extent children understand the math they're doing or if they are just following a series of rote procedures or steps. Good questions will allow children to clarify, generalize, and apply their math understanding. As my son Duncan said when he was ten years old, "The first thing to remember about math is that it should make sense. The second thing to remember is you can use it to order pizza."

"PAUSE"ATIVE BOX

Good Questions to Gauge Understanding

Take a moment to pause and think about how the following questions might help you assess your child's understanding of math. What other questions might be helpful?

Ask children . . .

- How did you use a picture, numbers, or words to show your thinking?
- How would you explain this problem to a younger child?
- What questions came up while you were working?
- What were you thinking as you selected strategies to solve the problem?
- What was the toughest part of the problem and why?
- Can you think of another way to solve the problem that might be quicker or easier? Explain.

(Adapted from Ontario Ministry of Education 2011.)

16. Sometimes I feel like pulling my hair out. What can I do to remain calm during homework time?

Homework time can test the most patient parent's resolve. If you feel yourself becoming overwhelmed, take a short break while children work independently for a while. It might be that you are hovering unnecessarily. Try giving children a little space and see if they complete the homework on their own. If you want to stay close by, use this time to do a little of your own math. Balance your checkbook or do a Sudoku or KenKen puzzle while children are working. This simultaneously reinforces the idea that math is an important real-world skill while keeping you from getting bored, impatient, or restless.

17. I think it would be easier for everybody if I just did my children's math homework for them. How can I resist the urge to take over their learning?

Research shows that parent involvement can have either a positive or negative impact on the value of homework. Your involvement can speed up a child's learning, engage you in the school process, and enhance your appreciation for education. Being involved in homework also gives you a chance to express support to your child and build positive attitudes about the value of success in school. But parent involvement may also interfere with learning, particularly if parents complete tasks that children are capable of carrying out independently.

Think about how you handle daily chores in your home. There are tasks that you complete almost without thinking—folding clothes, removing chewing gum from the carpet, mopping up spilled ice cream from the heating ducts, ordering takeout, and so on. But for children these things may be unfamiliar and difficult. Though it is quicker and easier to load the dishwasher myself, I know that doing so removes a learning experience from my children. When I take the time to teach my children *how* to load the dishes correctly, I'm preparing them for future kitchen success. Think of math homework as a chance to give children valuable skills for future math achievement. Children are not likely to be resourceful, persistent, or confident if you take over their learning. Give children enough freedom to succeed or fail—they'll learn from both.

18. I want to show my child how to do something the easy way or the way I learned how to do it. Is this all right?

There is often more than one way to solve a math problem. While some approaches are more efficient than others, no single strategy is more correct than another if you get the "right answer." Encourage children to use different methods to solve problems, such as by drawing diagrams or making an organized list. Ask children to explain how they solved a problem so that you can share in the thought process. By talking about a math problem, you and your child may find other ways to solve it.

One of my former fourth-grade students was raised by parents who attended school outside the United States. The student learned several unique, elegant approaches from her parents. When she shared these techniques with our math class, my students and I agreed that we can learn math from a variety of perspectives and using many creative approaches and still get a correct answer.

For more insights on problem solving, see the following questions addressed in Chapter 2:

Question No. 14, *Why is there such an emphasis on problem solving?*

Question No. 15, *What makes a person a good problem solver?*

Question No. 16, *How is my child's teacher helping my child learn to be a good problem solver?*

Question No. 17, *What questions can children ask themselves to help them make sense of and solve a tough math problem?*

19. How do I help my child when I don't remember how to solve the problem?

Don't let math amnesia slow you down. If you're like me, things you learned as a child have been pushed out of your long-term memory, replaced by show tune lyrics, quips from sitcoms, and sports statistics. I am able to recall every lyric from Les Misérables, dialogue from Seinfeld seasons 1–200, and the perky cheers I learned as a junior high pompom girl, but I must admit I have difficulty recalling the periodic table, which I once had memorized fully.

Admit to your child that it's been many years since you had the chance to use your formal math brain muscle. Refresh your memory by browsing through your child's textbook or other materials. Visit websites such as the National Council of Teachers of Mathematics' Illuminations resource, Khan Academy, or others that offer explanations, vocabulary, and hints for solving the rich problems your child encounters during homework time. And remember, it's OK to learn alongside your child. It's never too late to understand mathematics and gain an appreciation for its wonder and practicality.

There's More Than One Strategy

TEACHING TIP

You may have felt frustrated at being required to teach a specific strategy for a certain problem type. For example, the objective may state that students should use a number line to solve two-digit addition problems, but you feel the standard algorithm is quicker. After all, it's the strategy you learned and perfected as a child, right? But teaching a variety of computation strategies is essential to helping children build number sense, discernment, and confidence in mathematics. Consider how painters choose from a variety of brushes. Does the job need a brush with stiff bristles or supple ones? Wide or narrow? Flat, round, or fan-shaped? Synthetic or natural hairs? When we teach, value, and accept a variety of computation strategies, we communicate the idea that flexibility and efficiency are as important as accuracy in the grand scheme of mathematics. A painter limited to one style of brush will soon become frustrated at not achieving a quality result, and a student who relies solely on a standard algorithm misses the elegance, order, and creativity inherent in mathematics.

20. What should I do with completed and graded math papers that my child brings home?

Don't automatically throw your child's math papers away. Reviewing graded papers rather than immediately tossing them shows your child that math matters and that you are interested in your child's daily efforts, not just report card grades. Math papers can also tell you a lot about your child's math learning. When looking over papers brought home from school, listen carefully as children explain how they solved a problem. If an answer is incorrect, take this as an opportunity to show a math-positive mindset by learning from mistakes. The first thing to do is ask your child how the problem was solved. Work a few similar problems together. If possible, make corrections on the paper and return it to the teacher to show that you've both made an effort to correct errors and misunderstandings. You may choose to set the assignment aside to review again later (especially if emotions are getting in the way of communication). Be optimistic! Help children have a "can-do" attitude by praising their efforts as well as their accomplishments (Mirra 2004). Show children that math work is not just something to be completed and forgotten about. Learning from mistakes and celebrating accomplishments, even after a grade is set, builds a math-positive mindset in your home.

21. My child and I are both really frustrated, and I feel like our math homework time is damaging our parent-child relationship. At what point do I throw in the towel and hire a math tutor?

As a parent, I can empathize with your struggle. Sometimes my interaction with my son Duncan during homework is more *Clash of the Titans* than *Leave It to Beaver*. If you feel that working closely with your child on homework is doing more harm than good, but your child still needs one-on-one assistance, perhaps a private tutor or group tutoring situation would be a wise option. Commercial tutoring companies provide a variety of services, from targeted homework help to general study skills and work habits. Private tutors offer personal attention that can facilitate homework and preparation for exams. Even if you choose to take advantage of a professional tutor, I hope you'll still consider yourself your child's first and best cheerleader for math. You're not throwing in the towel from your role as supporter and encourager.

22. Where can I find a math tutor? What is a reasonable fee to pay?

A good resource for finding a math tutor for an elementary-aged child or middle schooler is the local high school's math club. If you live near a college or university, their math departments might also keep a list of math students who are qualified tutors. Searching online for math tutors is an option, but as with any service, ask for references and check them. Never leave your child unattended with a person you do not know well. Arrange to have the tutor provide services at a local library, café, or other public place if this is allowed. Check local service providers to find out what fees are reasonable for your area. Consider having your child pay a portion of the tutoring fee if you think that will help give them a nudge to work hard during tutoring time.

23. What should I expect or request from a math tutor?

You, your child, and the tutor should discuss your expectations up front. Do you want the tutor to focus only on completing homework assignments or on facilitating understanding of concepts in a general sense? What are your child's goals—better scores, better understanding, better motivation in math, or a combination of all three? How much time do you plan to devote to tutoring? How often and for how long will tutoring sessions take place? Once the tutor, your child, and you agree on all the logistics and expectations and tutoring sessions begin, reevaluate often to be sure that the tutor is a good fit for your child's learning approach and personality. Ask your child how the tutoring sessions are going. Watch to see if your child's math grade gradually improves and understanding grows. If, after several sessions, you don't see improvement, or you sense a negative attitude in your child, considering moving on to another tutor.

24. I don't want my children to think of tutoring as punishment or some indication that they aren't smart. How can I explain that tutoring is a way of helping them understand concepts rather than just "getting by" in math?

More than any other subject, math connects prior understanding to build new knowledge. If children have gaps in understanding, they may be able to muddle through for now, but those gaps crop up later and can cause difficulty in learning new content. Encourage children by saying that you are willing to help them at home or by hiring a tutor because you believe hard work pays off. Learning math thoroughly takes effort and dedication, but it is worth it!

25. What online math assistance is available for my child and for me?

Many parents and children use the Internet as a handy, free resource for math assistance. Consider the following:

- As a parent, you might choose to clear the cobwebs away by reviewing a specific type of problem online prior to giving help to your child.

- You might find the website hosted by your child's math textbook publisher a useful reference. Your child's teacher may give a password to fully access the resources.
- Search online to find tips for specific types of problems as well as video tutorials. See, for example, the math tutorial videos on Khan Academy's site.
- Several university mathematics education programs upload videos with explanations for mathematical modeling strategies or problem types. The University of Houston's YouTube channel offers more than forty short videos with explanations for solving problems using various mathematical models. Search YouTube using UHouston Math.
- Some commercial tutoring companies and private tutors also offer online tutoring services. These can be free, available via paid subscription, or pay-as-you-go formats.
- Dynamic learning programs and games provide adequate challenge, keep track of children's progress, and provide variety that helps engage and motivate kids. See the websites for Math Playground and Cool Math for Kids.

TEACHING TIP

Communicating with Parents about Tutoring

If you believe a student would benefit from tutoring for remediation or acceleration, approach parents tactfully and with professionalism. A quick informal email sometimes does the trick, especially if the tutoring would enhance a passion for mathematics. For remediation, be sure to offer your help in addition to suggesting other resources available commercially or through the school. Emphasize that tutoring in mathematics can make the difference between "just getting by" and deeply understanding content. Reassure parents that you believe their child has the capacity for high achievement in mathematics if given the right support and extending maximum effort.

26. My children love to read. I'm wondering if I can get them excited about math through reading. What are some good picture books that reinforce some of the math concepts my children are learning at school?

Pairing mathematics with a familiar and much-loved activity—sharing a picture book—can be a great way to encourage your child to do more math. Books with mathematical themes support mathematics learning in a comfortable setting—the shared reading experience (Ginsburg 2016). Some picture books use a math theme to tell a story. Math-themed picture books offer myriad possibilities for mathematical investigations (Monroe and Young 2018). But don't feel limited to picture books; familiar fairy tales can offer fun and teachable moments for math. For example, the gingerbread man met a cow and a fox. After reading the story, ask your child, which one of these animals is larger? Have them draw a picture showing their answer. The teachable math moments in the story of *The Three Billy Goats Gruff* might involve counting, comparisons of sizes, and ordinal numbers. After reading *Little Miss Muffet*, give your child the following task: "The spider frightened Miss Muffet, and she dropped her bowl of curds. She had only eaten half the bowl. Draw a picture to show how much she had left." Think of your own questions and activities for your child's favorite fairy tales and watch the math come to life in a variety of child-friendly contexts. Any time you can sneak in a bit of math with your shared reading experience, do it.

Tips for Selecting and Using Picture Books with Math Themes

Following are guidelines I use when selecting and using picture books for math learning. Take a moment to pause and think about these guidelines. Select a book and walk through these guidelines with it. Use the book with your child to build connections between reading and math.

- *Be Choosy.* Select books that your child and you find appealing or fun. Ditch boring stories, even if you think they may be "educational" (Ginsburg 2016).
- *Where's the Math?* Find the math in the story. Think of math-related questions. For example, you might wonder aloud, "There are a lot of animal friends who want to ride in the boat. I bet they weigh a lot. I wonder if the boat will be able to hold all of them?"
- *Make Connections.* Help your child connect the math to something they already know: "Have you ever tried to get a boat to float in the tub or the swimming pool? What if you put other toys on top of the boat? What happens then?"
- *Discuss the Math.* Talk about the math with your child. Model correct mathematical vocabulary; however, allow for informal language too. For example, you might say, "The animals in this story are trying to fit on a tiny boat. Do you think there is room? How much capacity do you think that boat has? Can it hold all of their weight?" Ask open-ended questions and encourage children to justify their thinking. For example, ask, "The boat already held an elephant and a giraffe. Why does the boat start to sink only when the mouse climbs aboard?"

"PAUSE"ATIVE BOX

Children's Literature for Math Learning

Following is a list of my favorite books for use with children's math learning. Take a moment to pause and review the list. Some of them may be classics you remember from your childhood. Others are new. Select a book and activity to try with your child. Alternatively, think of one of your child's favorite books; are there math-related questions and/or activities that would go with it? For additional recommendations, search online for the Mathical Book Prize for award-winning fiction and nonfiction math-themed books for kids of all ages.

Book	Grades	Summary	Math Topic	Math Activity
Kang, Anna. *You Are Not Small*, illus. Christopher Weyant. Seattle, WA: Two Lions, 2014.	PreK–2	Two creatures contemplate their size in relation to one another and to a larger group of creatures.	Measurement (comparisons)	Have children look around the room; what items can they find that are bigger than they are? What items are smaller? Are there more items that are bigger or smaller? Why do you think that is?
Atinuke. *Baby Goes to Market*, illus. Angela Brooksbank. Somerville, MA: Candlewick, 2017.	PreK–2	Unbeknownst to mama, baby puts her own yummy foods into the basket.	Counting	Play grocery store in your own kitchen. Make a shopping list with items you have on hand. Write a number next to the items. Have your child fill the order with the correct quantity for each item. If your child does not read yet, use pictures or symbols in place of words.
Lobel, Arnold. *Frog and Toad Are Friends*. New York: Harper and Row, 1970.	PreK–2	Frog has lost his button. He and Toad find many buttons but none of them are quite right.	Geometry (attributes)	Have children look closely at the buttons on their clothing. How are their buttons similar to or different from Frog's button?

(continued)

(continued)

Book	Grades	Summary	Math Topic	Math Activity
Berkes, Marianne. *Over in the Grasslands: On an African Savanna*, illus. Jill Dubin. Nevada City, CA: Dawn Publications, 2016.	PreK–2	This book counts the many animals that call the savanna home. A repeating pattern and rhyming make the book a fun read-along.	Data analysis (gathering data)	Make a frequency table (using tally marks), pictograph (using pictures), or bar graph (using colored-in grid paper) to record the number of each type of animal. Make sure to check for the hidden creatures on each page, too. Talk about which animals are the most common, the least common, and other interesting things you notice in the data.
Briére-Haquet, Alice. *One Very Big Bear*, illus. Olivier Philipponneau. New York: Abrams Appleseed, 2016.	PreK–2	Animals combine to make themselves as tall as Bear. He comes out on top, however, by gobbling up the teasing sardines.	Measurement (length, area)	Have children use sets of objects as nonstandard measuring units. For example: How many forks does it take to extend across the table (length)? How many napkins does it take to cover the table (area)? How many rolls of paper towels or toilet paper must be stacked to reach the top of the table (height)?
Hutchins, Hazel. *A Second Is a Hiccup*, illus. Kady MacDonald Denton. New York: Arthur A. Levine Books, 2007.	PreK–2	Time is explained in child-friendly ways: a second is a hiccup or the time it takes to turn around. A week is seven wakes and seven sleeps. In a month, a scraped knee has new skin.	Measurement (time)	Have children estimate how many times they can write their full name in one minute. Then try it out. Talk about how the estimate compared to the actual number of times they wrote their name.

(continued)

(continued)

Book	Grades	Summary	Math Topic	Math Activity
Daemicke, Songju Ma. *Cao Chong Weighs an Elephant*, illus. Christina Wald. New York: Arbordale, 2017.	PreK–2	Six-year-old Cao cleverly uses water displacement to weigh an elephant in ancient China.	Measurement (weight)	Use the Internet or informational books to research the weight of some of your children's favorite stuffed animals. Order the animals from heaviest to lightest.
Robinson, Michelle. *How to Wash a Woolly Mammoth*, illus. Kate Hindley. New York: Henry Holt, 2013.	PreK–2	This book outlines ten steps to a sparkling clean woolly mammoth.	Counting (ordinal numbers)	Read the story, having children replace Step 1 with *first*, Step 2 with *second*, and so on. Then have children use ordinal numbers to describe a familiar step-by-step process like brushing their teeth or getting ready for school.
Harris, Robie. *Now What? A Math Tale*, illus. Chris Chatterton. Somerville, MA: Candlewick Press, 2019.	PreK–K	Puppy uses wooden blocks to make a naptime bed.	Geometry (shapes)	Have children use wooden blocks or cutouts to compose new shapes. How can they combine the shapes to make a rectangular bed for a stuffed animal?
Messner, Kate. *Up in the Garden and Down in the Dirt*, illus. Christopher Neal Silas. San Francisco: Chronicle Books, 2017.	PreK–2	A girl and her nana plant a garden, opening the reader's eyes to the goings-on above and below the soil.	Geometry (positional words)	Have children go on a scavenger hunt to find out what is: (a) *above* the refrigerator, (b) *under* the couch, (c) *beside* the toaster, (d) *inside* the top drawer in the kitchen, and (e) *outside* of the front door. At the last stop you could leave a surprise such as a treat or book to be read.

(continued)

(continued)

Book	Grades	Summary	Math Topic	Math Activity
Gehl, Laura. *One Big Pair of Underwear*, illus. Tom Litchenheld. New York: Scholastic, 2015.	PreK–2	Can two bears share one pair of underwear? Silly counting, up-and-back patterns, and a host of helpful animals show how *counting everyone in* makes everyone happy.	Counting back, adding ten, place value	Roll a die. Have children say the number that is one less than the number shown on the die. This quick game helps build place value and counting skills. For an extra challenge, use a ten-sided die. For an extra-extra challenge, roll the ten-sided die, add ten to the number shown, and tell what number is one less. For example, if 8 is rolled, add 10 (18) and subtract 1 (17).
Myller, Rolf. *How Big Is a Foot?* New York: Young Yearling Books, 1991.	2–4	The king orders a bed as a gift for the queen, but no one knows how long it should be.	Measurement (need for standard units)	Compare the lengths of children's feet and yours. Discuss the importance of standardization of units. Make up or write a funny story using a unit other than feet (degrees, cents, minutes) to show what might happen if units were not standardized.
Neuschwander, Cindy. *Amanda Bean's Amazing Dream: A Mathematical Story*, illus. Liz Woodruff. New York: Scholastic, 1998.	2–4	Amanda learns how handy skip counting and multiplication are for everyday tasks.	Number and operations (multiplication)	Have children inventory the items in a cabinet or drawer by skip counting.

(continued)

(continued)

Book	Grades	Summary	Math Topic	Math Activity
Schoonmaker, Elizabeth. *Square Cat*. New York: Aladdin, 2011.	3–5	Eula, a square cat, feels out of place in a world of donuts and hoop earrings. However, she soon learns that being different is not necessarily a bad thing.	Geometry (shapes)	Ask, "What if there were no circles?" Children can draw a picture or describe what bicycles or cars would look like in a world of no circles.
Scieszka, John. *Math Curse*, illus. Lane Smith. New York: Viking Press, 1995.	3–5	Hilarious connections to math and the real world result when a child starts to see everything as a math problem.	Real-world math	Challenge children to tell about all the math-related experiences occurring during a typical day. Make your own mental list and see who comes up with the most math in action.
Viorst, Judith. *Alexander, Who Used to be Rich Last Sunday*, illus. Ray Cruz. New York: Silver Burdett, 1997.	3–5	Alexander's grandparents give him a dollar, but he frivolously spends it on candy and snake rental.	Number and operations (money)	Ask children, "Do you think Alexander spent his money wisely? What would you spend money on?"
Neuschwander, Cindy. *Sir Cumference and the First Round Table: A Math Adventure*, illus. Wayne Geehan. Watertown, MA: Charlesbridge, 1997.	4–6	Sir Cumference and Lady Di of Ameter must choose a table shape for their regular meetings with the Knights. They consider rectangles, ovals, and even octagonal shapes until the couple's son, Radius, points out that a fallen tree provides the ideal table top.	Geometry (shapes)	Explore the *Circle Tool* on the National Council of Teachers of Mathematics' *Illuminations* website. Adjust the size of the circle to show how the radius, diameter, and circumference are proportional to one another.

(continued)

(continued)

Book	Grades	Summary	Math Topic	Math Activity
Neuschwander, Cindy. *Sir Cumference and the Dragon of Pi: A Math Adventure*, illus. Wayne Geehan. Watertown, MA: Charlesbridge, 1999.	4–6	Clever Radius must solve a math riddle about circles to turn his father from a fire breathing dragon back into Sir Cumference.	Geometry (concept of pi)	Have children pick three different drinking glasses from the kitchen. Help children use a measuring tape to measure the circumference and diameter of the circular faces on the glasses, then work together to divide the two measures. The circumference divided by the diameter should be about 3.14. Try it with at least three more circular faces. What do you notice? This is a fun way to "discover" pi.
Birch, David. *The King's Chessboard*, illus. Devis Grebu. London: Puffin Books, 1993.	4–6	The king offers to reward a kind deed with a grain of rice for each square on a chessboard, doubling the previous day's offering, however finds his storehouses are quickly emptied.	Number and operations (multiplication)	Have children act out the first few pages of the story using a chess or checkerboard and coins. Have them guess which day will be filled with a coin value greater than one dollar. Talk about the doubling sequence and make an estimate of the total value of coins used on the board. Here's a hint: It's more than a million dollars!
Adler, David A. *Place Value*, illus. Edward Miller. New York: Holiday House, 2017.	4–6	Silly monkeys help the reader see why a digit's position in a number determines the digit's value. Learn some history of our counting system, a bit about decimals, and a lot about place value.	Counting, place value, large numbers, decimals	Take turns making up silly math sentences that show how mixing up the digits in a number can cause confusion and laughter. For example, *Miguel saved up $992 to buy a video game system. Molly has 61 stuffed animals on her bed. Mei read about an activist who walked backward 91 miles to raise money for charity.*

TEACHING TIP

Take-Home Book Bags

Take-home book bags are a good way to get more math and more reading into your students' homes. Set up a check-out system that works for your students and you. Choose a high-quality mathematics picture book and simple hands-on activities that require few materials so families can easily do the activity after reading the book. For example, the take-home book bag for *A Chair for My Mother* (Williams 1982) might include discussion prompts like:

- "What obstacles does the girl have to overcome to earn the money to buy her mother a chair?"
- "When have you saved your money to buy something?"
- "How do you think the girl felt at the end of the book?"
- "Why is it important but challenging to save money?"

For an activity, suggest the game *Dollar Dice*. Players need coins and a die. The numbers on the die correlate to a coin value: 1 for a penny, 2 for a nickel, 3 for a dime, and 4 for a quarter. If a 5 is rolled, it serves as a wild roll (any coin value), and 6 is a lost turn. Players take turns rolling the die and adding coins. The winner is the first player to reach exactly one dollar.

27. How do I respond when my child asks, "When am I ever going to use this?"

Children often want to know *why* they have to learn something, and as adults we need to help them see the relevance in the mathematics they are doing. My son Duncan once declared that he saw no use for two-digit multiplication and division using decimals. While I quickly found a connection to his monthly allowance, weekly chores, and a lengthy list of expenditures at local eateries, not all mathematical connections are that easy to make. However, all of the content in elementary school mathematics is relevant in daily life, whether we recognize it or not. Number and computation, fractions, shapes, algebraic reasoning, data analysis, and probability allow us to budget, share, complete puzzles, arrive on time, follow trends, and lose the lottery.

Some of the rigorous high school mathematics content may seem isolated from real-world applications, but abstract reasoning, perseverance, and work ethic build personal character and habits of mind that pay off in multiple contexts. Even if students never again use a particular math concept, learning is a journey and not a destination. Math concepts are cumulative. A seemingly irrelevant math idea may be the doorway to something children will use in a future math class or in their career. And even if they don't apply that particular math concept again, learning a new way to think is never a bad thing.

"PAUSE"ATIVE BOX

Math in the Workplace

Some people think that only engineers and scientists use math in their work, but all professions involve math to some degree. Take a moment to pause and think about the following examples. Share them with your children or students. What other professions can you add to the list?

- *Pharmacists* use sorting and proportions to dispense the correct amount of medicine to people.
- *Mechanics* use problem solving to figure out what is wrong with a car. They also use measurement to calibrate engine parts and computation to figure estimates.
- *Store managers* need counting skills to keep track of inventory, employee work schedules, and wages. Cashiers use counting skills to make change.
- *Carpenters and architects* use estimating and measuring skills when designing or building things. They also need to be able to visualize how shapes fit together.
- *Cooks* use estimating, measuring, and time management skills to prepare meals.
- *Floor and carpet installers* use estimating, measuring, and multiplication skills to figure out the size of the room and how much material to order. (Pagni 2000)

28. What are some resources that are especially helpful for making connections to math and the real world?

The Internet provides rich resources for children to explore math in the real world. Check out the following websites and share them with children to investigate how math is useful in a variety of intriguing settings.

Website	Summary	What to Check Out
www.thefutureschannel.com	The Futures Channel produces mini-documentaries highlighting the math, science, engineering, and technology found in a variety of occupations. A few videos are free. Others are available with a monthly subscription.	"Calculating the Power of the Wind" shows how giant windmills create electricity with blades, rotors, and wind. In "The Rhythm Track" a drummer uses fractions and the value of notes to create rhythms.
http://ed.ted.com/series/math-in-real-life	Ted Talks are searchable by topic, and math is a popular one. Find dozens of educational videos that show historical, futuristic, and practical applications for math.	Australian comedian and radio personality Adam Spencer riffs on prime numbers. Eduardo Sáenz de Cabezón responds to stereotypical fixed mindset beliefs about the purposes of math with humorous, poignant connections to science. (In Spanish with English subtitles)
www.thirteen.org/get-the-math	Get the Math is a multimedia project about math (algebra, in particular) in the real world of fashion, music, sports, video games, and more. It's geared toward teens, but preteens and young children will find the videos appealing as well.	Fashion designer Chloe Dao uses math to create a clothing design that retails for less than $35. Special effects expert Jeremy Chernick uses math to design lighting for filming explosions.

29. What outings might help my child see math in action?

Math isn't just for the kitchen table or the classroom. Take children on a math adventure to see it in an environment that may surprise them. Taking a math-themed field trip may not sound that exciting, but if the math venture is tailored to children's interests (a theme park, a favorite restaurant, a horse ranch, a car dealership, a veterinarian's office, or an ice cream factory) it can bring math to life in a personal way.

Any of children's favorite haunts offer glimpses of mathematics in action. One of my favorite Mom-and-Duncan dates was to the used bookstore. My son Duncan loves to read and has a few favorite book shops where he delights in grabbing a stack of books and crashing on the store's coffee-stained couch for a couple of hours—days, if I'd let him. So, after a particularly challenging homework session where Duncan employed all the math-positive mindset he could muster, I promised him a trip to the bookstore as a reward. He decided to take along a few books to sell to the shop. In the car, I asked Duncan if he thought the math he'd been studying, decimals and percentages, offered practical uses. Unable to conjure up any useful purposes, Duncan seemed skeptical when I told him that the bookshop might be just the spot to see math in action. When we arrived at the shop, Duncan approached the sellers' desk with his wares. The staff member gave him a ticket and told him to browse, then return to the sellers' desk in fifteen minutes to receive an offer for his books. Duncan found roughly thirty-seven books about the history of the fast-food industry and lugged his plastic basket of trivia to the sellers' desk to find out if providence shone on him. The staff member handed Duncan a sheet of paper: $0.79 for each paperback book and $1.13 for each hardcover book *if* Duncan sold his books outright to the store. If Duncan chose to receive a store credit that he could redeem immediately, each paperback was worth $0.89 and hardcover books were worth $1.25. *Voila!* Math in action in a setting and purpose that Duncan found meaningful.

TEACHING TIP

Math-Related Field Trips

If you have input in a class field trip, consider a math-themed adventure. Even a mini field trip to a local grocery store can open students' eyes to real-world mathematics essential to everyday life. A sense of spatial relations is needed for stocking canned goods, an understanding of degrees to record temperature readings in the cooler and freezer sections, knowledge of dates to avoid spoilage in the dairy section, an understanding of scales to properly weigh produce, and, of course, financial literacy to be able to perform monetary transactions at the checkout. All these actions show how mathematics solves problems and makes our lives more fulfilling. Conclude the grocery store tour with a stop at the bakery for a pastry and insight into how bakers estimate the number of goods to bake, measure ingredients, and adjust cooking times and temperatures.

30. How can I take advantage of those teachable moments outside of homework time?

Look for them! Teachable moments in math occur naturally throughout a child's day; there are an abundance of unplanned opportunities that help children understand that math is all around them in their math-focused home. Setting the table for dinner involves patterns: *fork, plate, knife, spoon, fork, plate, knife, spoon*. Baking a batch of cookies involves measurement, fractions, and the occasional problem-solving scenario: *If we need to add $\frac{3}{4}$ cup of butter but only have a $\frac{1}{2}$ cup measuring cup, what can we do? How much dough did we eat before we baked it? One-third? One-half?* A trip to the park might include counting squirrels, seagulls, and roly-poly bugs. Figuring out the time difference before calling grandma in another time zone gives real-world experience with elapsed time. Watching the gas pump numbers fly by shows how quickly money can be spent.

Teachable moments might also come when children see that *you* use mathematics in your daily life. Point out the math involved in making your grocery list, shopping online for bargains, comparing bids for painters after your child gets creative with a permanent marker, and so on. Share your real-world mathematics experiences by talking out loud about the process of balancing a checkbook or tripling a chili recipe for the annual cook-off. Even if children are too young to understand the details, they will see and hear you taking time to work through ideas, checking your reasoning, finding and correcting mistakes, and using mathematics to solve important problems.

Teachable moments help boost children's opinion of math as a tool for living. My son Duncan often accompanies me to the grocery store. Teaching him to figure the unit price for items helps him see math as a useful skill. Using coupons for name brand items may or may not save money over a generic brand. Every day, people face situations that involve mathematics, such as reading a map, choosing the shortest route to a destination, setting a schedule, or determining the price of an item on sale. Help children realize that mathematics is a significant part of everyday living and a tool for piloting a successful life.

Writing a grocery list shows real-world applications for math, gives practice with writing the digits, and voices opinions about snacks!

Thinking about how coupons work gives children practice with saving and spending.

After a trip to the park, my children sometimes write about the math they saw. Two spiders have 16 legs!

31. How can I initiate family conversations about math?

Math isn't like the birds and the bees. Don't keep it secret and don't wait to talk about it until children are older. Research shows that talking about math is positively related to preschool-age children's early mathematical ability. Even after controlling factors such as a mother's education level and overall time spent talking during the day, talking about math made the biggest difference in math skills (Susperreguy and Davis-Kean 2016).

You don't need a math PhD; you just need to be chatty. In our math-focused family, we've cultivated a culture of openness and interest in math by talking about it during dinner. In today's hectic world, it can be a challenge to schedule family meals. But, at the very least, turn off the television, phones, and other devices during meals and interact with your children. Frequent family meals have been linked to lower rates of obesity, eating disorders, depression, school absenteeism, and substance abuse and to higher grades and reports of high-quality parent-child relationships (National Center on Addiction and Substance Abuse at Columbia University 2012).

Make the dinner table prime math-talk time with positive, light conversations. This isn't the time to grill your child or bad-mouth math, and it doesn't have to be forced or awkward. Instead of asking, "How was math today?" (a prompt that generally leads to a one-word response of "OK"), spark the conversation with questions like:

- What can you teach us about the math you're learning at school?
- What questions do you have about what you're learning?
- Can you tell us about a really tough math problem you worked hard to solve?
- Have you played any fun math games? Can you tell us about them?

QUICK REFERENCE CHART

Common Concerns about Learning Math at Home

This table is a quick reference for some of the *most common concerns* I hear from parents. The table lists the concern, an explanation of the thinking behind it, and an idea that addresses the concern. Consider photocopying this table and placing it somewhere where you can be reminded of what is needed to build your math-positive mindset and that of your children and/or students.

Concern	Explanation	What to Do	Question Number(s) (to Learn More)
Why homework?	Homework reinforces skills and concepts, gives time to finish assignments, connects parents to school, and teaches responsibility and good study habits.	Provide time, space, and age-appropriate help as needed to show that you value homework, too.	1
Homework time expectations	The right amount of math homework can foster independence, responsibility, and time management.	Strive for ten minutes of homework for each grade level (10 min. for first grade, 20 min. for second, etc.).	2
Difficulty level of homework	Teachers generally assign the same homework to all their students, even though we all realize students have individual strengths and weaknesses.	Having open communication with your child's teacher makes it easier to express concerns about homework being too easy or too difficult. Teachers may have some ideas for enrichment or be willing to ease up on the requirement.	5

(continued)

(continued)

Concern	Explanation	What to Do	Question Number(s) (to Learn More)
Why are homework routines important?	Routines help children know what to expect so they can get right down to work.	Set up a distraction-free space where children can complete homework before other after-school activities, if possible.	3, 4, 6
Should I have math tools on hand?	Having a homework space stocked with handy math tools (manipulatives) will limit scrambling to find them when needed.	Provide pencils, pens, ruler, calculator (when appropriate), and manipulatives to make mathematics homework a hands-on event.	7, 8, 10
Is money a math tool?	Yes! Use coins to practice skip-counting, build abstract thinking, and reach financial literacy goals.	Play *Coin Trading Game* together (see page 83).	8
How might virtual manipulatives help?	Base ten blocks, geometric shapes, spinners, and many more virtual manipulatives can provide a semiconcrete experience with math content.	There are several online resources for virtual manipulatives. Most math textbook publishers offer free online manipulatives you can access at home. Ask your child's teacher for an online access code.	9
Keeping homework stress free	Positive, stress-free homework sessions support positive associations with math.	Play math games, take brain breaks, and use movement (wiggle breaks) to lighten the mood.	11
A child shuts down during homework	Some frustration is normal, but tears and anger can interfere with positive learning.	Rule out emotional and physical causes, begin with easier homework problems, work for 10–15 minutes then take a brain break, and encourage persistence.	12, 13

(continued)

(continued)

Concern	Explanation	What to Do	Question Number(s) (to Learn More)
Asking good questions	Questioning is a good way to determine to what extent children understand the math concepts or if they are just following a series of rote procedures or steps to solve a problem.	Ask children to show and explain their thinking. What other ways can they think of to solve the problem?	14, 15
Staying calm	Math-positive parents set a calm tone during homework, offering support without being overbearing.	Don't hover unnecessarily. Step away when you feel yourself becoming anxious.	16
Helping without taking over	Telling children what to do is never as effective as letting them figure things out on their own, even if the latter takes more time (and teeth gritting). When we take over the thinking, we undermine children's confidence.	Support math-positive persistence by encouraging children to try multiple strategies before stepping in. When children feel supported, they are empowered to persevere.	17
Is it OK to show the way I learned it?	We may be partial to the algorithms and shortcuts we learned in school. Your child's teacher will probably introduce these *after* learning the concepts.	Remember there are many right ways to approach a math problem. Sharing your experience and having your child teach you a few newfangled strategies might help you both gain an appreciation for the diversity of approaches.	18

(continued)

(continued)

Concern	Explanation	What to Do	Question Number(s) (to Learn More)
I don't remember this math!	Things you learned as a child have been pushed out of your long-term memory.	Refresh your memory by browsing through your child's textbook or other materials. Visit websites such as the National Council of Teachers of Mathematics' Illuminations resource, and Khan Academy. Learn with your child!	19
What do I do with all these graded math papers?	Graded math papers can tell you a lot about children's competency; don't throw them away thoughtlessly.	Look over papers brought home from school; ask children to explain how they solved a problem. Praise their efforts. Show children that math work is not just something to be completed and forgotten about.	20
Hiring a math tutor	It's better to understand mathematics than to "just get by." Hiring a tutor may also improve your relationship with your child if math homework is causing conflict.	Get recommendations from friends or school. Check with the local university to find out if their math lab is for school-age kids. Interview several tutors to find a good personality and philosophy fit. Let your child know tutoring is not a punishment but a chance to really understand mathematics deeply.	21, 22, 23, 24

(continued)

(continued)

Concern	Explanation	What to Do	Question Number(s) (to Learn More)
Help! I need assistance.	Assistance in math can also be found online via video tutorials, virtual manipulatives, games, tutoring services, and more.	Visit the websites of the following organizations: • National Council of Teachers of Mathematics • National Library of Virtual Manipulatives • Khan Academy • Math Playground • Cool Math for Kids	25
How do reading and math support each other?	Encourage more reading, encourage more math! Math and reading are mutually supported when we share books with a math-positive theme.	Turn reading time into math time by pointing out the math in fairy tales, picture books, informational texts, and stories emulating math-positive mindsets. Search online for the Mathical Book Prize for award-winning fiction and nonfiction math-themed books for kids of all ages.	26
When am I ever going to use this?	All of the content in elementary school mathematics is relevant in daily life, whether we recognize it or not.	Point out that while some skills may seem isolated or impractical, math concepts are cumulative. A seemingly irrelevant math idea may be the doorway to something children will use in a future math class or career.	27

(continued)

(continued)

Concern	Explanation	What to Do	Question Number(s) (to Learn More)
Where's math in my world?	Mathematics is a way of making sense of the world—a language, a pattern, an order. When we recognize and appreciate the mathematics around us, we are open to learning more about its potential for humankind.	Help children put on their math-seeing glasses. Use online videos to open children's eyes to the relevance of math in careers, hobbies, and society. Take a math-focused field trip to the grocery store, beach, or your own place of business. Math is only thinly disguised by our own lack of awareness.	28, 29, 30
Why talk about math?	Research shows that talking about math is positively related to preschool-age children's early mathematical ability. Even after controlling factors such as a mother's education level and overall time spent talking during the day, talking about math made the biggest difference in math skills.	You don't need a math PhD; you just need to be chatty. Cultivate a culture of openness and interest in math by encouraging conversations about it at the dinner table and more.	31

I Can Help with Number and Operations

CHAPTER 4

(continued)

(continued)

1. How can I support my preschool-age child in counting?

We all work diligently at teaching our children to read; however, you may be surprised to learn that mastering early mathematics skills predicts later school success even more than reading ability does (Duncan et al. 2007). Helping children with math can start at an early age; support preschoolers in becoming competent counters by taking advantage of everyday opportunities for meaningful counting. For example, children need a lot of practice with the counting sequence. Count together while washing hands, walking up the stairs, or cleaning up toys. If you have a few extra minutes, sit on the floor across from your child and roll a ball back and forth, counting as you go. Don't worry if your child makes a mistake. For a long time, my son Duncan skipped the number 7 when counting. With a math-positive mindset, we view these mistakes as chances for the brain to grow. Correct the error and move on, giving preschoolers' brains those neural connections that will make them successful in future math learning.

"PAUSE"ATIVE BOX

What Does Counting Involve?

The early skill of counting may seem simple but is actually a fairly complex cognitive exercise. Take a moment to pause and think about what the skill of counting requires. Read the following table with your child in mind—what does your child need the most practice with? Try out the corresponding suggested activities with your child.

Counting involves . . .	Math Activities
• saying numbers in the correct sequence	Practice counting with your child as you climb steps or empty the dishwasher.
• keeping track of what's been counted	Counting books, like *Ten Little Ladybugs* (Gerth 2000), are great for keeping track of what's been counted. Have children place a counter on top of each object as they count. That way, they won't miss anything or count anything twice.

(continued)

(continued)

Counting involves . . .	Math Activities
• recognizing that the number sequence is consistent no matter what is counted or in what order the objects are counted	Count a set of objects from left to right then from right to left. Talk about how it doesn't matter which direction we go; the counting process and number order is the same.
• using one-to-one correspondence (assigning one number to each object counted)	Have children touch each object as they count. If children say several numbers at once, place your hand gently on the child's, bouncing up a bit between touches. Or you can have the child pick up objects one by one and move them to the other side of the table as they count. Either of these rhythmic movements paces their counting and helps them maintain one-to-one correspondence.
• understanding cardinality (recognizing that the final number spoken is the quantity in the set)	When children finish counting a set, ask, "So how many do you have?" If the child can't answer, you can tell them. "You have four." Or ask the child to go and tell someone in another room how many they have. The child is unlikely to say, "I have 1, 2, 3, 4." Instead, the child will say, "I have 4." This makes the connection that the final number is the one that's really important.

2. What are the big ideas for kindergarten counting and how can I help my kindergarten-age child learn to count?

Kindergarten-age children build on the counting skills they developed in their preschool years, including one-to-one correspondence, subitizing, and cardinality. Many standards emphasize counting in kindergarten as an important prerequisite to addition and subtraction. There are a lot of things you can do to help children develop these skills. From having children count spoons as they set the table to pointing out the digits on the speed limit sign, you are in a perfect position to support counting and numeral awareness.

"PAUSE"ATIVE BOX

Summary of Big Ideas for Kindergarten Counting

Here are some of the big ideas for counting kindergarten-age children will learn in school, including activities you can do to support their learning. Pause and take a moment to review the list. Put an asterisk next to some of the activities you'd like to try. Try them with your child. How did it go?

Big Ideas for Kindergarten Counting	Math Activities
Count to 100 by ones and by tens.	Counting by 10s is efficient for large sets. Make a trail mix using ten of each ingredient: raisins, dry cereal, baked crackers, pretzel sticks, chocolate candies, and so on, to total 100 items. Then count by 10s to find the total.
Begin counting from a number other than one.	Model this for children by saying, "I see we have three, four, five napkins on the table" or "There are four, five, six toys to put away." It will sound less odd to children if they hear you start counting on in this way.
Write numbers from 0 to 20 to show how many objects are in a set.	Have children write grocery lists that include quantities or package sizes. For example, 2 bananas, 4 pounds of grapes, and 12 rolls of toilet paper.
Connect counting to cardinality (the last number given tells the number in the set).	After children finish counting a set, ask how many are in the set. If they cannot tell you without recounting the whole set, you can work on cardinality by having the child put the objects in a cup as they count. Then ask, "How many did you put in the cup?" This draws a connection between the counting process and the total number in the cup.
When counting objects, pair each object with one and only one number.	This is called one-to-one correspondence. Sometimes children rush when counting. Ask children to count slowly and loudly. For some reason, loud counting is more rhythmic and helps children say just one number per item counted. Strange but true!
Understand that the number of objects is the same regardless of their arrangement or the order in which they were counted.	Put a handful of dry beans or cereal pieces in a cup. Shake and spill them onto the table. Have children count them, place them back in the cup and repeat several times. Talk about how the arrangement of the objects and the order they are counted doesn't change the number of objects in the cup.

(continued)

(continued)

Big Ideas for Kindergarten Counting	Math Activities
Understand that each successive number name refers to a quantity that is one larger.	Use two colors of blocks, counters, or dry beans. Start building the number 5 by setting five blocks of one color in a vertical line. Place the sixth piece at the top of the line using a new color. Continue building to 20, showing how the number grows by one each time.
Count to answer "how many?" questions of up to 20 objects in various arrangements (scattered, lined up, etc.).	Count as you do chores such as laundry. Count the clothes as you fill the washer. Count the clothes again as you fill the dryer. The number stays the same even though the clothes are in a different arrangement. (No guarantees that they'll be the same number at the end the drying cycle, however. Where do those missing socks go?)
Create sets of up to 20 objects.	Gather collections to count at the park or playground. Have your child gather leaves, sticks, bugs, rocks, and other items and count them. Counting is fun when we take it outside.
Identify whether the number of objects in one group is greater than, less than, or equal to the number of objects in another group.	Have family members make a block tower with one block for each letter in their name. For example, my children Will, Sybil, Duncan, Chas, Zeb, McGregor, Quinn, and Knox would take 4, 5, 6, 4, 3, 8, 5, and 4 blocks, respectively. Compare the towers to see whose are tallest (McGregor), shortest (Zeb), and equal (Will, Chas, and Knox, for example). This activity develops numerical and length comparisons.
Compare two written numbers between 1 and 10.	Write three numbers on a card. For example, 4, 10, and 7. Use real-world contexts to bring meaning to greater and less than. For example, to practice *greater*, ask, "Which of these numbers would you prefer for presents on your birthday?" Then say, "Yes, ten is greater than four and greater than seven." For practicing *less than*, you could ask, "Which of these numbers would you prefer for chores on Saturday?" This time, rather than making the comparison, allow your child to supply the reason for their choice. *Greater* and *less than* make more sense when we situate them within a child's experience.

3. I heard a new term, *subitizing*. What is this?

I'm so glad you asked. Subitizing was the topic of my doctoral dissertation. I fell in love with this fascinating topic through my study and could talk about it for hours on end. Just ask my family! The topic is truly fascinating and definitely worthy of study.

Subitizing is the immediate, accurate apprehension of the number of items in a small set—usually up to three or four items for young children. It's basically "seeing" the quantity without having to count each item separately. Imagine that you're walking across your child's cluttered bedroom holding a handful of marbles. You step on a pointy block and, cringing, accidentally drop the marbles onto the floor. As you pick them up, you notice that you don't have to count them; you "see" a set of four marbles, three marbles, and, aha, here is another set of three marbles.

Though the term *subitizing* is only beginning to gain popularity in the elementary math curriculum, it is not a new concept. In fact, the marble scenario I just described is a nod to Scottish philosopher Sir William Hamilton, who first documented the phenomenon in 1859. In 1871 another fellow, Jevons, informally tested the idea by grabbing a handful of dry beans and throwing them into a box, glancing at them briefly and estimating the quantity. He then counted the beans to check if his subitizing was accurate. After completing 1,027 trials, Jevons noted that sets of up to four beans could be estimated with complete accuracy and sets of five beans were estimated with very few errors. Pretty cool, eh?

Remember that perceptual subitizing is only possible with a small set of items. For sets of more than four items arranged in a random display, counting is required. But larger sets arranged in a familiar pattern, such as those on dice or dominoes, can be subitized quickly because they are quickly recognized. We call this spatial subitizing. You can help children practice spatial subitizing by playing dominoes and dice games with them. Encourage them to not count but to quickly subitize the lower quantities and recognize spatial arrangements for numbers larger than four. How would you subitize or "chunk" the dots on this card?

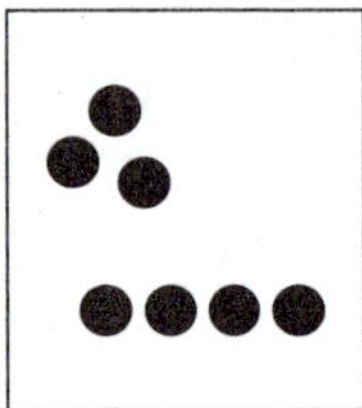

I'm guessing you saw a triangular-shaped 3-dot formation on top and a row of 4 on bottom. You were able to use perceptual subitizing for those small sets of 3 or 4 dots. Then you use conceptual subitizing to chunk them and join them to find the total number of dots—7.

The final type of subitizing relates to chunking small sets and combining them. Called conceptual subitizing, this process is a foundation for the counting on strategy children will use when they begin adding.

"PAUSE"ATIVE BOX

Clean-Up Subitizing Game

Take a moment to pause and think about how everyday chores in your math-focused home might contribute to subitizing. For example, clean-up time can be math time; as children are cleaning up toys, have them grab a handful of blocks or other small toys and quickly subitize the number. Grab a handful and say, "I picked up four. This time I got three. Here are two." Children will soon join in the clean-up subitizing fun. The results? A tidy room and strong subitizing skills.

TEACHING TIP

Playing Subitizing Games

You can reinforce spatial and conceptual subitizing by playing ten-frame games. Ten-frames are rectangles with two rows of five boxes and can be found from many sources online and in books, including *It Makes Sense! Using Ten-Frames to Build Number Sense* by Melissa Conklin (Math Solutions, 2010). Make a deck of ten-frames and laminate them for longer use. Try these three games in math workshop learning stations and more.

- *Game 1:* Divide a deck of ten-frame cards. Each player turns over their top card and sees who can name their number the quickest. For the ten-frame that follows, the answer is 8.
- *Game 2:* Turn over the card and add 1 or 2 to the amount shown. For the ten-frame that follows, the answer would be 9 or 10.
- *Game 3:* Instead of saying how many boxes are filled, say how many are empty. For the ten-frame that follows, the answer would be 2.

Have fun practicing subitizing and building number sense with your children.

●	●	●	●	●
●	●	●		

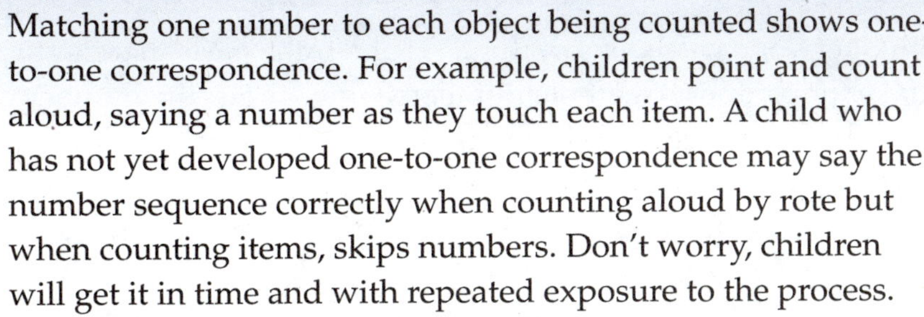

4. Sometimes my children skip numbers when they are counting things. Why does this happen?

Matching one number to each object being counted shows one-to-one correspondence. For example, children point and count aloud, saying a number as they touch each item. A child who has not yet developed one-to-one correspondence may say the number sequence correctly when counting aloud by rote but when counting items, skips numbers. Don't worry, children will get it in time and with repeated exposure to the process.

You can help children develop one-to-one correspondence by having them pick up and move items from one side of the table to the other, counting each as they do so. This simple task requires children to slow down and be more intentional about matching each number spoken to each item counted. You can also help children develop one-to-one correspondence by gently putting your hand on their hand and helping them point and count slowly. Then have children do it again by themselves, touching each item as they count aloud. Encourage children to count slowly and clearly. When children are told to count slowly and clearly, they often establish a rhythm that assists in the one-to-one correspondence. Even asking them to count loudly may work, if you can tolerate the noise.

5. My child is in kindergarten and sometimes writes numbers backward. Why is this?

This is a normal stage of development where children reverse the number as they are writing it. Don't be alarmed and don't overreact. This phenomenon (also very common in writing letters such as lowercase *b* and *d*) usually disappears on its own by second grade. Simply show your child the correct orientation for the number. It might be helpful to post numbers (especially those your child is reversing) around the house so they are seen often. For instance, you might write the number on an index card and hang it on your child's bathroom mirror or next to the front door.

6. A homework assignment asks my child to solve using number bonds. What is a number bond?

Number bonds are ways of thinking about the parts that might make (or compose) a whole number. For example, the number 7 can be composed of four different sets of number bonds: 6 and 1; 5 and 2; 3 and 4; and 0 and 7. These are the number pairs that compose the whole number 7. Number composition and decomposition help children to think flexibly about numbers—to pull them apart and put them back together again in useful ways. This part–part–whole thinking is powerful for computation, reasoning about how numbers relate to each other, and building number sense. For example, when a child is adding 25 + 7, it is useful to be able to break that 7 into 5 and 2. The 5 is easily added to the 25 to make 30 and then the 2 can be added to make 32.

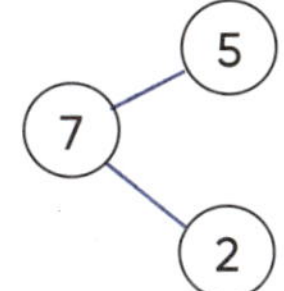

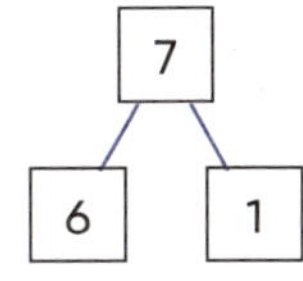

Illustrating two of the four sets of number bonds for the number 7: 5 and 2, 6 and 1.

"PAUSE"ATIVE BOX

Number Bonds

Knowing number bonds helps solve computation items with multiple digits. Breaking apart numbers makes the addition easier. Take a moment to pause and think about the equation, 25 + 17 = _____. What are some ways you might solve this by thinking about number bonds? Once you have at least one way in mind, review the following ways—is your way one of them?

Strategy 1

Break apart one number:

25 + 17 =

5 12

30 + 12 = 42

(continued)

(continued)

Strategy 2

Decompose the numbers using 10 as a "friendly" number that's easy to compute with. Break apart the number 17 to 7 and 10. Then add 10 to the 25: 25 + 10 = 35. Finally add 7 to 35 for a sum of 42.

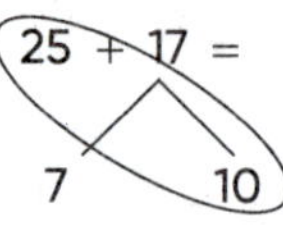

35 + 7 = 42

Strategy 3

If it helps to compute, break it down even further using sets of five. Counting by fives is easy, right? 25, 30, 35, 40. Then add the 2 for a sum of 42.

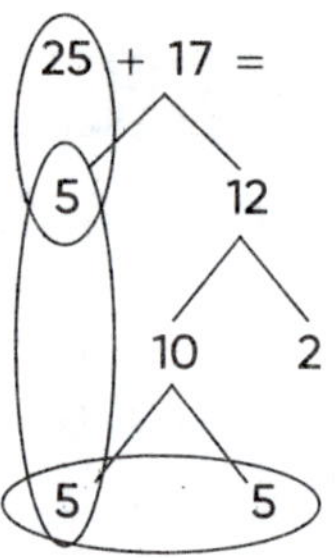

25 + 5 + 5 + 5 = 40

40 + 2 = 42

7. Beyond counting, what else does my child need to be able to do with numbers?

Number is a huge topic! Number provides the foundation for all other content areas in math—geometry, measurement and data, and algebra. Consequently, it is the focal point of much of elementary school mathematics. Number (that's not a typo—there is no "s" on the end) refers to all things related to our glorious numbering system. Number includes counting, of course, but also judging the reasonableness of our computations, estimating, using mental math, place value, and much more. Number is like a jam-packed trail mix—all its components combining to make a wholesome, practical energizer that gets us through when the going gets tough. Number includes understanding how addition and subtraction relate to one another and how multiplication and division work. Number includes acquiring basic facts and understanding the effects of the commutative, associative, identity, and distributive properties and how they help us develop fluency with operations. Number takes in fractions, decimals, percentages, ratios, and proportions. Number also involves appreciating how all these things work in harmony to keep our trail mix filled with enough peanuts, sunflower seeds, raisins, and chocolate chunks so we can move along in our math venture.

"PAUSE"ATIVE BOX

Summary of Big Ideas for Number and Operations for Kindergarten through Grade 5

Here are some of the big ideas related to number and operations for study in each grade (National Governors Association Center for Best Practices and Council of Chief State School Officers 2010). Take a moment to pause and go through this list with your child in mind; place an asterisk next to the ideas you'd like to try. Try them with your child. How did it go?

Grade	Big Ideas in Number and Operations	Math Activities
K	Compose and decompose numbers from 11 to 19 into tens and ones using objects, drawings, and equations.	Play *Heart Broken Numbers*. Cut out 10 hearts and write a number from 11–19 on back of each. Fold each heart in half and write one addend on one of the blank sides of the heart. Lay the heart, addend side face down, on the table. Have children use counters to compose the whole number, then turn the heart over, break off the addend shown and count to find the missing addend.
1	Understand a two-digit number is composed of tens and ones. • Think of 10 as a bundle of ten ones called a "ten." • Think of 11 to 19 as "a ten and some ones." • Think of the decade numbers (10, 20, 30, etc.) as sets of 10 and 0 ones. • Compare two-digit numbers using >, =, and < symbols.	Have children play the *And 10 More* game. Children roll a 10-sided die and read the number shown. Then say, "And 10 more makes. . . . " For example, if 9 is rolled, say, "9 and 10 more makes 19." Have children do a Choral Count (Franke, Kazemi, and Turrou 2018) to practice adding 10 starting with any number. It might sound like this: 8, 18, 28, 38, 48, 58, 68, 78, 88, 98. Write down the numbers as children say them and look for patterns.
	Use manipulatives, drawings, place value, properties of operations, and/or the relationship between addition and subtraction to: • Add within 100. Add a two-digit number and a one-digit number, and a two-digit number and a multiple of 10. • Subtract multiples of 10 in the range 10–90 from multiples of 10 in the range 10–90.	
	Mentally find 10 more or 10 less than a number without having to count; explain the reasoning used.	

(continued)

(continued)

Grade	Big Ideas in Number and Operations	Math Activities
2	Understand a three-digit number is composed of hundreds, tens, and ones. • Think of 100 as a bundle of ten tens called a "hundred." • Think of 100, 200, 300, etc., as "sets of hundreds and 0 tens and 0 ones."	Play *Base Ten Riddles* (Van de Walle, Karp, and Bay-Williams 2019) to practice place value. Have children build the number with base ten blocks, draw, or write as you give clues. Here are a few to get you started, then work together to make up some new riddles. • I have 23 ones and 4 tens. Who am I? (63) • I have 4 hundreds, 12 tens, and 6 ones. Who am I? (526) • I have 30 ones and 30 hundreds. Who am I? (3,030) • I am 450. I have 250 ones. How many tens do I have? (20) • If you put 30 more tens with me, I would be 1015. Who am I? (715)
	Explain why addition and subtraction strategies work, using place value and the properties of operations.	
	Numbers within 1,000: • Skip-count by 5s, 10s, and 100s. • Read and write using numerals, number names, and expanded form (e.g., 900 + 60 + 5 = 965) • Compare using >, =, and < symbols. • Mentally add or subtract 10 or 100 to a given number, 100 through 900. • Add and subtract using manipulatives, drawings, place value, properties of operations, and/or the relationship between addition and subtraction.	
	Numbers within 100: • Fluently add and subtract using place value, properties of operations, and/or the relationship between addition and subtraction. • Add up to four two-digit numbers using place value and properties of operations.	
3	Use place value to round whole numbers to the nearest 10 or 100.	Discuss the concept of rounding with children. Rounding is helpful when we use it to change a tricky number to a number that's easier to compute with. For instance, rounding 198 to 200. Here's the general rule for rounding: If the number you are rounding is followed by 5, 6, 7, 8, or 9, round the number *up*. Example: 46 rounded to the nearest ten is 50. If the number you are rounding is followed by 0, 1, 2, 3, or 4, round the number *down*. Example: 233 rounded to the nearest hundred is 200.
	Fluently add and subtract within 1,000 using place value, properties of operations, and/or the relationship between addition and subtraction.	
	Multiply one-digit whole numbers by multiples of 10 using strategies based on place value and properties of operations.	

(continued)

(continued)

<table>
<tr><th>Grade</th><th>Big Ideas in Number and Operations</th><th>Math Activities</th></tr>
<tr><td rowspan="2">4</td><td>Multidigit numbers:
• Recognize that a digit in one place represents 10 times what it represents in the place to its right.
• Read and write using numerals, number names, and expanded form.
• Compare using >, =, and < symbols.
• Use place value to round to any place.
• Fluently add and subtract using the standard algorithm.</td><td rowspan="2">Using partial products helps make computing with large numbers easier.
For example, we can solve $23 \times 37 =$ by breaking apart the numbers to make them easier to multiply.
$20 \times 30 = 600$
$20 \times 7 = 140$
$3 \times 30 = 90$
$3 \times 7 = 21$
Then we add those up.
$600 + 140 + 90 + 21 = 851$
Try a few problems with children and see how place value understanding helps in computing with large numbers.</td></tr>
<tr><td>Use equations, rectangular arrays, and area models to explain and illustrate how to use place value and/or the properties of operations to:
• Multiply a whole number of up to four digits by a one-digit whole number, and multiply two two-digit numbers.
• Divide using four-digit dividends and one-digit divisors.</td></tr>
<tr><td>5</td><td>Multidigit whole numbers:
• Recognize that a digit in one place represents 10 times as much as it represents in the place to its right and $\frac{1}{10}$ of what it represents in the place to its left.
• Multiply whole numbers by powers of 10 and explain patterns in the number of zeros in the product. Use exponents to denote powers of 10.
• Fluently multiply using the standard algorithm.
• Use equations, rectangular arrays, and area models to explain and illustrate how to use place value and/or the properties of operations to divide with four-digit dividends and two-digit divisors.</td><td>Saying decimals correctly supports children's understanding. Instead of saying "three point five" use the correct terminology, "3 and 5 tenths." Doing so helps children connect decimals (a new idea for fifth graders) to a concept they've been working with for several years—fractions.
Practice with children by saying the following:
$\frac{2}{10}$
0.2
$\frac{75}{100}$
0.75
$\frac{115}{100}$
1.15</td></tr>
</table>

(continued)

(continued)

Grade	Big Ideas in Number and Operations	Math Activities
5	Decimals: • Multiply and divide decimals by powers of ten and explain patterns in the placement of the decimal point. • Read and write decimals to thousandths using numerals, number names, and expanded form. • Compare decimals to thousandths using >, =, and < symbols. • Use place value understanding to round decimals to any place. • Add, subtract, multiply, and divide decimals to hundredths, using concrete models or drawings, place value, properties of operations, and/or the relationship between addition and subtraction; relate the strategy to a written method and explain the reasoning used.	Notice that the word we use to represent the decimal point is *and*. So, the number 1.15 should be said "one and fifteen-hundredths." To be consistent in the use of the word *and* to signify the decimal, we should say the number 115 as "one hundred fifteen" instead of "one hundred and fifteen."

8. What is number sense?

Teachers and math experts sometimes throw around terms like *number sense* and expect that everyone will know what it means. However, guess what? Number sense is a much easier term to use than it is to define. Number sense is a "big idea" in math that includes estimation, place value, basic facts, relative size of numbers, counting strategies, and number composition and decomposition. It makes math experts feel smart to use all those big words, but what does it mean? We say people have strong number sense when they:

- understand how our base ten number system functions,
- can make logical estimates,
- can compute fluently,
- have a good idea of how numbers compare to one another in magnitude (relative size of numbers),

- can count in flexible ways such as counting on from a larger number, and
- can pull numbers apart and put them together (number composition and decomposition).

Number sense is difficult to define partly because it involves creativity and flexibility with numbers. It involves finding shortcuts (hey, that sounds like a Mathematical Practice—see Chapter 2, Question No. 22, *My state adopted the Common Core State Standards for Mathematics (CCSSM). What do I need to know about the CCSSM?*) and looking for sensible ways

"PAUSE"ATIVE BOX

A Number Sense Experiment

Take a moment to pause and do the following number sense experiment: *If you wanted to quickly subtract 52,997 from 53,505 without using a calculator, what would you do? Try it.*

Did you write the equation vertically and follow the traditional algorithm? Even without paper and pencil, some people find themselves setting up the standard algorithm in their minds and trying to solve it that way. That set of steps is strongly engrained in us after years of practice. But is it the most efficient approach for this problem? A strong sense of number opens up more possibilities. Consider the following two strategies.

Strategy 1

Decompose the difference into the part above 53,000 (505) and the part below 53,000 (3).

$$505 + 3 = 508.$$
$$53{,}508 - 53{,}000 = 508$$

Strategy 2

Use **compensation**, adding 3 to the 52,997 to make it 53,000, and adding 3 to 53,505 making it 53,508. The difference between the two numbers stays the same, but the numbers are easier to work with.

$$53{,}508 - 53{,}000 = 508$$

What other mental math strategies might you use? There are many "rights"—meaning many *right ways* of solving this subtraction problem. And many of those right ways involve number sense.

to make arithmetic easier (that sounds like more Mathematical Practices). We might say that a person who looks for alternatives to time-consuming, memorized algorithms has a strong "number sense disposition."

When children develop number sense, they think flexibly about numbers, which increases fluency with counting and computation strategies and lays a foundation for mental math. Children also see some of the elegance, simplicity, and order of our number system. Number sense is understanding that our number system is beautiful and understanding what makes it so.

9. How do base ten blocks and a place-value chart help my child understand numbers?

In our numbering system, the position of the digit shows its value. A place-value chart helps children see that the 5 in the number 258 has a value of 50. The chart also helps with composition (putting together) and decomposition (pulling apart) of numbers. For example, the number 258 is the same as 2 hundreds, 5 tens, and 8 ones but can also be thought of as 25 tens and 8 ones.

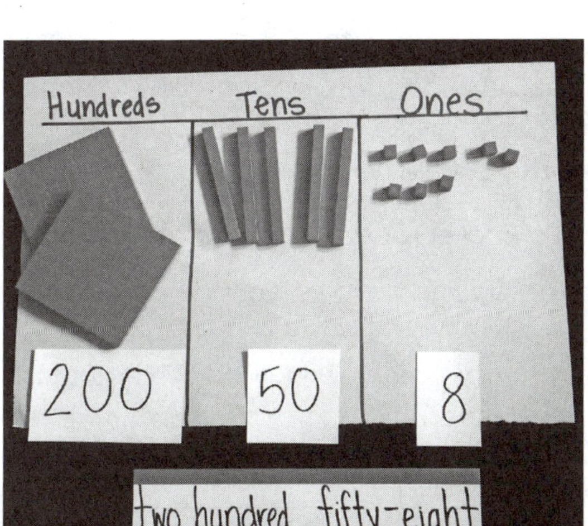

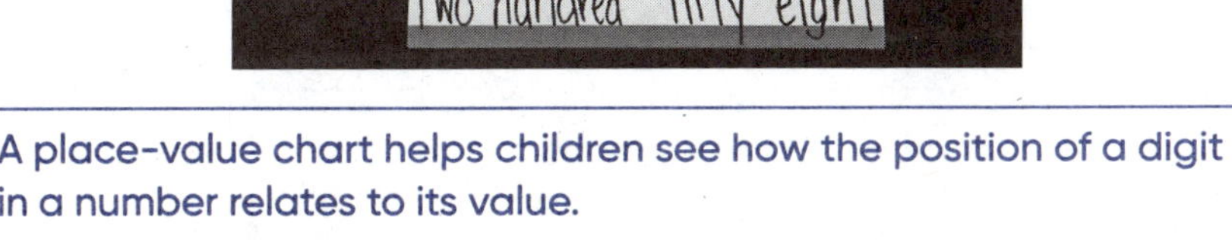

A place-value chart helps children see how the position of a digit in a number relates to its value.

In the place-value chart shown here, you see columns for hundreds, tens, and ones. The pieces in the picture are manipulatives called base ten blocks. Sometimes we use technical terms for the blocks. They come in hundreds called flats, tens called rods, and ones called units. There is also a thousand cube not shown in this picture. (And stored in my garage is a ten thousand rod made from ten of those thousand-cubes hot-glued together. Cool, eh? A bit bulky, but a fun visual for how our base ten numbering system follows a pattern.) The place-value chart and base ten blocks are powerful concept-building partners to help children visualize relationships in our base ten system.

TEACHING TIP

Using the Hundreds Chart

If you are using the calendar as a daily routine in your classroom, you might consider shortening up the time spent on the calendar and adding a hundreds chart talk and exploration. Why? Our numbering system is base ten. The calendar is base seven. Not really all that helpful to students working to become fluent with tens and ones. Young children typically know the 1–9 sequence and a bit beyond but often have difficulty identifying the counting patterns for decades (10, 20, 30) and those pesky transitions (for example, that 39 signals 40 is next) (Baroody and Wilkins 1999). The calendar does not give any counting practice beyond the number 31. Here are a few powerful ways to interact with a pocket-chart-style hundreds chart.

Ideas for Using a Pocket-Chart-Style Hundreds Chart

- Mix up the cards and have students "fix the chart." Discuss as appropriate.
- If students struggle with teens, build the chart just to 20. If desired, flip the teens numbers over to red to make them stand out. Ask questions like, "What number is *one larger* than 17?" "What number comes *before 15*?" "What patterns do you see as you look down the columns?" "How does looking at the row above help with the teen numbers?" Support this experience by having students make sets of fewer than

(continued)

(continued)

20 objects (buttons, counters, flat marbles) in a learning station, then having students show you how many objects there are by pointing to the correct number on the hundreds chart.

- Build the chart to 50 then show the multiples of 10 using cards of a different color (such as red). The multiples of 3 make a cool design on the board when they are built using a different color. So do the multiples of 11. If your hundreds chart doesn't come with different colors of cards, make your own. The patterns help students see the order and beauty of our number system.
- Ask students to identify a missing number.
- Ask students to complete a row or column on the chart. (The column one is tough!)
- Build the chart to 100. Play *Secret Number*. For example, tell students to start at 27. Go down one, go to the right one, go down two. Where did you land? (Fifty-eight.) Can students do it in their heads after a while?
- Play games where you say a number and students tell the number that is one more or two more, one less or two less.
- Make the hundreds chart one of your learning stations during math workshop so that more students can interact with the materials. Encourage students to use the red, purple, and green number cards that come with the pocket-style chart to show patterns they discover.

This is a pocket-chart-style hundreds chart I use for leading small- and whole-group explorations with our number system. I like it because the cards are removable. At the beginning of PreK or kindergarten, start by filling the chart to 20. Later, build to 30, then 50 and 100.

10. Back in my day, we called it carrying and borrowing. Is that what regrouping means?

Our number system doesn't change, but terms do. Understanding the meaning of and reasons for new terminology in math will help you be a math-positive parent. Regrouping is the process used in addition and subtraction that most of us remember as "carrying" and "borrowing." Math experts prefer the term *regrouping* because it matches the physical process of changing a number from one form to an equivalent form. Regrouping helps in doing arithmetic computations. For example:

$$\begin{array}{r} 67 \\ -\ 19 \\ \hline \end{array}$$

Pretend we did some regrouping with base ten blocks and a place-value chart. You would first place a set of six rods in the tens column and seven units in the ones column. Then to subtract, you would need to remove nine from the ones column. But wait, there are only seven ones. So you would regroup a set of tens by exchanging one rod for ten units and placing those in the ones column. For example:

$$\begin{array}{rl} 67 = & 50 + 17 \\ \underline{-\ 19} = & \underline{10 + \ \ 9} \\ & 40 + \ \ 8 = 48 \end{array} \qquad \begin{array}{r} 50 - 10 = 40 \\ 17 - 9 = \ \ 8 \\ 40 + 8 = 48 \end{array}$$

The shortcut way to solve the subtraction problem using one traditional algorithm looks like this:

$$\begin{array}{r} {\scriptstyle 5\ 17} \\ \not{6}\not{7} \\ -\ 19 \\ \hline 48 \end{array}$$

To add using regrouping, a problem such as 19 + 13 would require us to "carry" 1 from the ones place (because 9 + 3 = 12) to the tens place. Since this digit really represents the number 10, it's more appropriate to say we're regrouping it by changing those ten ones into one 10.

Communicating Math Language to Parents

TEACHING TIP

Many parents are confused by the language associated with reforms in mathematics. Consider yourself a bridge between the *old ways* and current approaches. Be careful with the use of confusing or unfamiliar terminology and help parents see how changes in mathematics education aren't really changes in math but in our ways of teaching children to be better, more flexible thinkers who can apply mathematics rather than just memorize steps. You are essential in the quest to build math-positive parents!

11. The word problem on my child's homework uses the term *about*. Does that mean the problem calls for estimation? What's that all about?

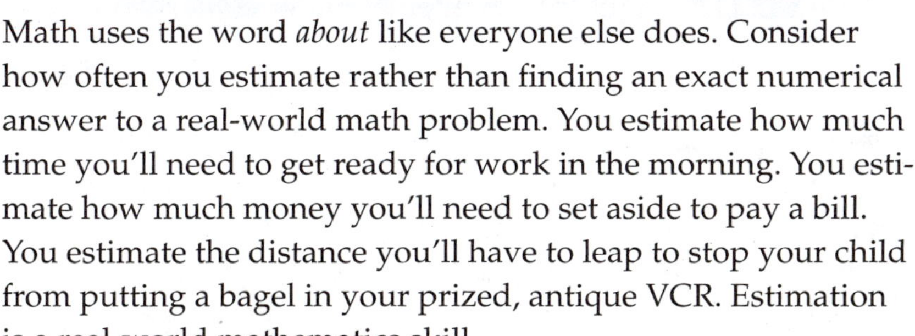

Math uses the word *about* like everyone else does. Consider how often you estimate rather than finding an exact numerical answer to a real-world math problem. You estimate how much time you'll need to get ready for work in the morning. You estimate how much money you'll need to set aside to pay a bill. You estimate the distance you'll have to leap to stop your child from putting a bagel in your prized, antique VCR. Estimation is a real-world mathematics skill.

In word problems, the word *about* is sometimes used to indicate that the answer will be an estimate rather than a precise number. About means "very close to" or "almost the same as" (Monroe 2006). For example:

> The twenty-four students in Mihir's class each have about twenty-one milk jug lids to donate to the recycling contest. About how many lids will the class donate to the contest?

Problems such as these allow children to use mental math or other strategies to make a quick estimate, and a math-positive mindset gives children the confidence to take a risk and make an informed guess.

12. What is mental math?

Don't let mental math mess with your head. Just use your head. Mental math requires solving a computation problem without using paper and pencil, a calculator, or manipulatives. It is a valuable mathematical skill to be able to compute quickly and accurately using the best math tool, your mind. Have a math-positive mindset, and don't be intimidated. You usually have a great calculator in your head. Mental math is not a series of magic tricks, but rather the solid understanding of number relationships and flexible thinking. With a little practice, anyone can learn calculation skills that save time and avoid the hassle of reaching for the calculator or paper and pencil. Doing a little bit of mental math regularly helps build efficiency and accuracy. The estimation we do every day depends on the mastery of some mental math strategies. The next few questions address some of these strategies. (To learn more about considerations for choosing paper and pencil, a calculator, manipulatives, or mental math refer to Chapter 3, Question No. 10, *What about using calculators? Isn't using a calculator cheating?*)

13. What can I do to help my children learn their addition facts?

Fluency with addition facts is foundational for many other areas in mathematics because it connects strongly to the other operations, place value, and number sense in general. Building fluency doesn't have to mean hours of flash cards or drill and kill worksheets. Instead, aim for a range of strategies for learning facts rather than memorizing them. Learning implies thinking and understanding, while memorizing is often fleeting and fragile.

It's likely your child's teacher has some specific aims related to addition facts. In kindergarten, the goals for addition relate to building the concept of addition as the joining of two sets to find a sum. Kindergarteners do a lot of hands-on work with manipulatives and word problems (stories about addition) to construct this idea with sums up to 10. Kindergarteners also work to understand composing and decomposing numbers and to build fluency for sums up to 5. In first grade, children are introduced to a range of number-sense rich strategies for sums of up to 20. Strategies like using known facts, number bonds for the number 10, number splitting, and *Bridging to a 10* are all highly useful for building fluency with facts. First graders strive for fluency for facts up to 10. For second grade, expect children to shift into high gear for attaining a firm understanding of fluency with facts up to 20. Automaticity or mastery of the addition facts is generally expected by the end of third grade.

"PAUSE"ATIVE BOX

Helping with Mental Math: Addition

Here are a few ideas for helping children become adept at mental math for addition. Children who can think flexibly about numbers, pulling them apart and putting them back together in many ways, can compute efficiently. Pause and take a moment to try out some of these strategies. You'll see how handy it is to be agile with arithmetic.

Reasoning Strategies for Mental Addition

Strategy	Example	Comments
Using Known Facts	*17 + 17 = ?* *I may not know 17 + 17, but I know 10 + 10 is 20 and 7 + 7 = 14. So 20 + 14 = 34.*	This strategy is sometimes called derived facts. It offers a springboard between known and unfamiliar facts, allowing children to apply something they know to help them figure out something they don't.
Number Bonds for the Number 10	*1 + 9 = 10* *2 + 8 = 10* *3 + 7 = 10* *4 + 6 = 10* *5 + 5 = 10* 10: 9 1 10: 8 2 10: 7 3 10: 6 4 10: 5 5	These five pairs of numbers are the only ways of getting 10 using two addends, and it is best for children to know them by heart. Doing this will speed up computation, because it facilitates the *Bridging to a 10* strategy shown later in this table and, when paired with place value understanding, can be applied to large numbers as well.

(continued)

(continued)

Strategy	Example	Comments
Number Splitting	*Look at 7 + 8.* *Imagine separating the 8 into 3 and 5.* *You can take those 3 and add them to the 7 to make 10.* *And that will leave the other 5 (of the 8) to add on to the 10 to make 15.* *So 7 + 8 becomes . . .* *7 + 3 + 5 which becomes . . .* *7 + 3 is one of those "ways of getting 10" that children know.* *10 + 5 = 15*	Number splitting allows you to pull numbers apart in ways that make the problem easier. Practice by using two empty ten-frames and beans. For the example shown here, place 7 beans on one ten-frame and 8 on the other. Then ask children how they could solve the problem by moving some beans from one ten-frame to the other. Taking 3 beans from the 8-bean frame and adding to the 7-bean frame fills it up. The child can easily see a full frame (10) and 5 more to find the answer.
Bridging to a 10	*Now let's look at a slightly larger addition item:* *Suppose you have to find 38 + 15.* *Instead of counting, use number splitting to bridge to a 10.* *So in the case of 38 + 15, add 2 to 38 to make a 10—40.* *Then add the other 13 (of the 15) to 40 to get 53.*	Just like we saw in the previous example, 10 is a handy anchor for adding. Getting to the nearest 10, often called *Bridging to a 10*, makes the subsequent jump easier. If it helps, you can suggest that children visualize ten-frames (as in the previous example). When I use this strategy, I often think of the nearest ten being a position on a number line. Whatever ways children visualize and apply this strategy, it's a powerful way to build fluency in computation.
Two-Digit Addition without Regrouping	*52 + 34 = ?* *First, look at the ones place. If the problem doesn't require regrouping, then you can first add the numbers in the tens place.* *50 + 30 = 80.* *Next add up the numbers in the ones place: 2 + 4 = 6.* *Finally add the 80 you got from adding the tens, and the 6 you got from the ones, to get 86 as the answer.*	A good sense of tens and ones (place value) helps children determine right away if regrouping is required.

(continued)

(continued)

Strategy	Example	Comments
Double-Digit Addition with Regrouping	*47 + 36 = ?* *Bridge to 10 by adding 47 + 3 = 50* *Then add 33 (that's 36 – 3)* *50 + 33 = 83*	Thinking of a number line helps here. 47 48 49 50 83 Three little jumps to bridge from 47 to 50 then a big jump of 33 to get a sum of 83. It's easy to add 50 + 33, isn't it?
Compensation	Example A *68 + 19 = ?* *Add 1 to 19 to make 20:* *19 + 1 = 20* *Then add 68 + 20 = 88.* *Subtract 1 from 68 to compensate:* *88 – 1 = 87* *So, 68 + 19 = 87* Example B *68 + 19 = ?* *Add 1 to 19 to make 20:* *19 + 1 = 20* *Compensate by subtracting 1 from 68:* *68 – 1 = 67* *Add 67 + 20 = 87* *So, 68 + 19 = 87*	To compensate, you can do either of the following: • Adjust one of the numbers and then adjust the answer (Example A). • Adjust both numbers. Then it's not necessary to adjust the answer (Example B). Understanding that numbers can be pulled apart and put back together in flexible ways makes compensation a useful strategy.

14. My children are doing okay with addition, but what can I do to help them learn their subtraction facts?

Let's talk about what you should "take away" from subtraction. Subtraction is given less emphasis in the memorization realm. Why? Well, because a child with strong addition fact fluency can apply that knowledge to subtraction problems as well. Subtraction is the inverse operation (the flip side) of addition.

"PAUSE"ATIVE BOX

Helping with Mental Math: Subtraction

Children may use subtraction reasoning strategies like those used in addition. Here are a few ideas for helping children become adept at mental math for subtraction. Pause and take a moment to try out some of these strategies.

Reasoning Strategies for Mental Subtraction

Strategy	Example	Comments
Using Addition	*27 – 19 can be thought of as 19 + ? = 27.* Or . . . *I know it's 8 because 9 + 8 = 17, then I add the other set of 10.*	This example shows how "thinking addition" can help solve subtraction problems.
Number Splitting	*27 – 19 might be thought of as 27 – 17 then – 2 more.*	Use knowledge of number composition and decomposition (breaking numbers into parts) to solve problems efficiently.
Using a Multiple of 10 as a Friendly Number	*27 – 19 = ?* *I think 20 is easier to work with than 19. So, I'm going to add 1 to the 19 to make it 20. 27 – 20 = 7* *Then I need to adjust the answer by adding to 7 to get 8.*	Similar to the addition strategy *Bridging to a 10*, this approach makes the problem easier to solve using mental math. In the example shown here, it eliminates the need for regrouping.

(continued)

(continued)

Strategy	Example	Comments
Compensating	Example A *46 – 19 = ?* *Add 1 to 19 to make 20:* *19 + 1 = 20* *Then subtract 20 from 46:* *46 – 20 = 26* *Add 1 to 26 to compensate: 26 + 1 = 27* *So, 46 – 19 = 27* Example B *46 – 19 = ?* *Add 1 to 19: 19 + 1 = 20* *Compensate by adding 1 to 46: 46 + 1 = 47* *Subtract 47 – 20 = 27*	To compensate, you can do either of the following: • Adjust one of the numbers and then adjust the answer (Example A). • Adjust both numbers. Then it's not necessary to adjust the answer (Example B). Just like with addition and place value, understanding that numbers can be pulled apart and put back together in flexible ways makes compensation a useful strategy.
Bridging Back to a 10	*28 – 11 = ?* *Take 8 away from 28 to get to 20.* *Then take 3 more away, leaving 17.*	Working backward with 10 as a "bridge" is preferred when the nearest 10 is less than the original number in the problem.

15. Why is my child learning so many different types of multiplication?

Really, there is only one multiplication operation. Children are not learning different *types* of multiplication; they are learning different *models* for multiplication. Looking at multiplication in different ways helps students see applications for multiplication in the real world. There are five basic models for multiplication, explained in the following table.

Multiplication Model	Description	Example(s)
Model 1: Repeated Addition	Thinking of multiplication as the same number added over and over.	2 × 3 can be said "two groups of three" and thought of as adding 3 two times: 3 + 3.
Model 2: Sets or Groups	Thinking of objects that come in sets.	Liam has 4 packages of chewing gum. Each package contains 6 sticks. He has 24 sticks of chewing gum to share with his sisters.
Model 3: Rectangular Array	Thinking of objects that are arranged in rows and columns where the objects are not touching. (A little mathematical vocabulary comes in handy when discussing the area and array models for multiplication. Remember that a row is horizontal, and a column is vertical. A convention in mathematics holds that when describing an array or area model, we list the rows first and then the columns.)	Cookies on a baking sheet, arranged in rows and columns, are a rectangular array.
Model 4: Area	The *area* model, like the rectangular array, uses rows and columns but the sections are connected. Using a row and column description, we could say this pan of brownies is a 5 × 4 area model for multiplication.	The area model is like a pan of brownies cut into pieces but still touching each other in the pan.

(continued)

(continued)

Multiplication Model	Description	Example(s)
Model 5: Combinations	Multiplication as *combinations* (sometimes called Cartesian products) involves finding all of the different ordered pairs for two or more sets.	You have three shirts and three pairs of pants to choose from. How many different outfits can you make? You have a choice of chocolate or vanilla ice cream and caramel or butterscotch topping. If you have just one scoop and one topping, how many different combinations can you make? (Pizza toppings and sandwich fixings also make great contexts for combinations problems.)

16. What are some specific strategies, besides memorization, that can help my child compute quickly using multiplication?

Just as children need to develop fluency with addition facts (and to a lesser degree subtraction), they need to be able to solve multiplication items flexibly, efficiently, and accurately using a variety of appropriate strategies (Parrish 2010, 2014). Math experts tell us that children go through three stages when learning basic math facts: (1) modeling with objects and/or counting, (2) using reasoning strategies, and (3) mastery by being able to efficiently produce answers (Van de Walle, Karp, and Bay-Williams 2019).

Stage 1: Modeling and/or Counting

First, children need ample time to develop the concept. This generally involves lots of hands-on experience with manipulatives and counting objects or skip counting with sets of objects to find the product.

Stage 2: Reasoning

The next stage, reasoning, is too often neglected and is crucial to a strong sense of number. This stage involves reasoning about how multiplication facts relate to one another and cluster together. Children also use known facts to derive answers to unknown facts. Experts say that when children use reasoning to "know" their facts, children rely less on memorizing and more on understanding and can apply their knowledge of basic facts to more advanced forms of computation.

Stage 3: Mastery

Mastery suggests a child has achieved fluency with the multiplication facts. This might mean the child knows the facts from memory. But "knowing from memory" is not necessarily the same thing as "memorizing." With practice (which can be embedded in rich problems or isolated fact-fluency experiences), children can begin to "just know" their multiplication facts. We call this mastery because children are applying reasoning strategies, known facts, and other strategies seamlessly—flexibly, efficiently, and accurately applying a variety of appropriate strategies to call up multiplication facts.

How does my family handle mastering the multiplication facts? By eliminating known facts, identifying facts that need practice, and working on it a little at a time. One afternoon when my son Duncan was in third grade, he brought home a gigantic stack of flash cards with multiplication items printed on them. Since he was expected to practice his facts from 0×0 up to 12×12, this stack contained 169 cards he had painstakingly cut apart during school! You can imagine his enthusiasm about this daunting task. Sure, he needed to be able to solve these items quickly; however, did he (and I—ugh!) need to spend hours and hours flipping through all those flash cards?

No!

Thankfully, understanding a few key ideas about multiplication eliminated most of the flash cards. For example, Duncan didn't need to drill those $0 \times n = 0$ cards; all he needed to understand was the Zero Property of Multiplication—the product of any number and zero is zero. With this understanding, Duncan tossed twenty-five cards in the trash! Next, we talked about the Identity Property of Multiplication. He explained to me that any time you multiply a number by 1 the product is the original number. So long to all twenty-three of the $1 \times n = n$ cards! (Remember, we'd already gleefully tossed the 0×1 and 1×0 cards.) Next, I set out pairs of cards with "turnaround" facts such as 3×5 and 5×3 and 2×9 and 9×2. I asked Duncan if he noticed anything interesting about these cards. He told me that you can multiply the numbers in either order and get the same result. I supplied the fancy mathematical term for this, the Commutative Property, and we tossed one from each pair of "turnaround facts." So long to about fifty-five cards. (For more about the properties of multiplication, see Chapter 7, Question No. 7, *What's the importance of the properties of the operations?*)

The stack of flash cards, now noticeably shorter, appeared much less intimidating to both of us. We were on a roll! Next, we took out the cards showing multiplication by 2, 5, and 10. These were "the easy ones" according to Duncan. He'd been counting by 2, 5, and 10 since kindergarten. He quickly showed mastery of those twenty-two cards and triumphantly removed them from the stack.

Now we could focus our time on the multiplication facts Duncan needed to practice. We went through the remaining thirty or so cards and found those he already knew well, setting those aside so that we could spend our energy on the less-familiar ones. Whew! You can see how when you help children "understand" rather than just "memorize" multiplication facts, it makes the task much less overwhelming.

Sequencing the Multiplication Facts

TEACHING TIP

Did you know there's a sequence for teaching multiplication facts to help children progress in their understanding, reasoning, and mastery? Kling and Bay-Williams (2015) recommend the following order:

1. *Foundational Facts*
   - 2s, 5s, 10s
   - 0s, 1s, and the multiplication squares (2 × 2, 3 × 3, etc.)
2. *Derived Facts*
   - Adding or subtracting a single group (I don't know 8 × 4, but I know 8 × 3 = 24, then I add 8 more.)
   - Halving and doubling (I don't know 9 × 12 but I know 9 × 6 = 54, then I double that to get 108.)
   - Using a square product (I'm good with squares. I can use 7 × 7 = 49 to find 8 × 7 by adding 49 + 7 to get 56.)
   - Decomposing a fact (I don't know 11 × 12, but I can break 12 into 10 and 2. I know 11 × 10 = 110 and 11 × 2 = 22. Then I can add them together to get 132.)

The foundational facts lay the groundwork for derived strategies. Derived strategies build on number sense and often emerge from children's thinking during number talks (Parrish 2010, 2014) and other contexts. Carefully sequencing your teaching, practicing, and expectations for mastery will help students build math-positive mindsets for multiplication facts and flexible understandings that can be applied to all sorts of computation problems.

"PAUSE"ATIVE BOX

Helping with Multiplication Facts

Here are a few ideas for helping children become adept at multiplication facts. Pause and take a moment to try out one of these strategies. Which ones make sense to you? What ideas do you have for making these strategies meaningful to children?

Reasoning Strategies for Multiplication Foundational (or Basic) Facts

Strategy	Example	Comments
Zero: $n \times 0 = 0$	Your grandma gave you 2 packages of gum. But you ate all the pieces of gum inside. You have 2 packages of 0 pieces of gum. How much gum do you have?	This is the *Zero Property of Multiplication* and children need a good bit of practice with it conceptually, so give a lot of examples before having them memorize this.
One: $n \times 1 = n$	1 package of gum containing 7 sticks of gum equals 7 sticks of gum.	This is the *Identity Property of Multiplication* and children need a good bit of practice with it conceptually but soon can apply it quickly as a learned fact.
Two: $n \times 2 = n + n$	There are 10 people in the Cutler family. If each person has 2 eyes, how many eyes does this make?	Children generally begin skip-counting by twos in first grade and can apply that rote practice when multiplying by 2. Also, multiplying by 2 builds on the doubles addition facts, which children generally commit to memory with greater ease than some of the other addition facts.
Five	There are 10 people in the Cutler family. If each person has 2 hands with 5 five fingers on each hand, how many fingers do the Cutlers have?	Counting by fives is generally introduced during kindergarten. Children practice the rote skip-counting sequence and can apply that knowledge when multiplying by 5.
Ten	There are 10 people in the Cutler family. If each person has 10 toes, how many toes do the Cutlers have?	Counting by tens is generally introduced during kindergarten. Children practice the rote skip-counting sequence and build on place value knowledge to multiply numbers by 10 quickly.

(continued)

(continued)

Strategy	Example	Comments
Doubles	To solve 14 × 8, use doubles. 14 × 2 = 28 28 × 2 = 56 Double 56 is 112.	Knowledge of doubles builds on skip-counting. I know 14 doubled is 28. Double that again to get 56. Double that again to get 112.
Double then Halve	To solve 25 × 12, double the 25 making it 50. 50 × 12 = 600. Half of 600 is 300.	This is a cool strategy! Here is an example. 16 × 5 Double the 5 to get 10. 16 × 10 = 160 Half of 160 is 80. Give it a try on your own: 13 × 5 7 × 15
Close Facts	To solve 11 × 11, solve a close fact. Think how 10 elevens are close. 11 × 10 = 110. Then add one more set of 11. 110 + 11 = 121.	Close facts can be larger or smaller numbers. Choose facts that are easier for you to use.

17. Why is my child learning so many different types of division?

Like multiplication, we look at division in different ways to help children understand the concept and make sense of division problems. We divide division up into measurement division and partitive division. The difference between these two types of division problems is in the unknown—what you are trying to solve for. (For more about different types of word problems, see Chapter 2, Question No. 14, *Why is there such an emphasis on problem solving?*)

Division Type	Description	Example(s)
Measurement Division	The number of *sets* is unknown.	Sybil has 60 colored pencils. She uses 12 pencils a week. How many weeks will her pencils last? Sybil has 15 coloring pages. If she draws on 3 pages a day, for how many days will her coloring pages last?
Partitive Division	The number of *items in each group* is unknown.	Sybil has 12 carrots that she wants to share with 3 friends and herself. How many carrots will each person get? Sybil has 33 coloring pages. She wants each of her 5 friends to have the same number of pages. How many coloring pages will each person get?

Knowing the different models for division problems helps children visualize what is actually going on during the division process. For the colored pencil problem shown as an example of measurement division in the above table, children might draw sixty lines to represent the pencils. Then children might loop a circle (representing a week) around a set of twelve pencils. That's for twelve pencils a week. Children would find all of the sets of twelve pencils and count the sets—five weeks with twelve colored pencils each week.

For the partitive division problem involving carrots (see the table on previous page), children might draw twelve carrots and then draw four little people to represent themselves and three friends. Children might draw lines from the carrots to the friends and say something like, "One for you, one for you, one for you, one for you. Two for you, two for you, two for you, two for you," and so on as the carrots are fairly shared with each friend. Here is what it might look like:

18. My children know their multiplication facts well. How do I help them practice division facts?

Rather than drilling the division facts in isolation, start by building the connection between multiplication and division in the same way you connect addition and subtraction as inverse operations (or flip sides). For example, to solve $56 \div 8$, think: *what number times 8 equals 56?* Do this, then apply the appropriate multiplication reasoning strategy. In this case, you might use a close fact: *I know that 5×8 is 40. I need 16 more to get to 56, so that's two sets of 8—$7 \times 8 = 56$.*

"PAUSE"ATIVE BOX

Helping with Division Facts

Here are two ideas for helping children become adept at division facts. Pause and take a moment to try out one of these ideas. Consider how you can help children connect multiplication and division in meaningful ways to facilitate these two strategies.

Reasoning Strategies for Division Facts

- *Strategy 1:* Help your child to see how a problem can be solved using either multiplication or division. For example: *Santiago has 16 trading cards in his collection. If he puts them in his binder that has pockets for 4 cards per page, how many pages have cards in them?* Help your child see that this problem can be solved two ways: 16 ÷ 4 = ? and 4 × ? = 16.
- *Strategy 2:* Think multiplication, then apply a multiplication reasoning fact (see the table, Reasoning Strategies for Multiplication Foundational [or Basic] Facts, on pages 160–161).

19. I thought math wasn't supposed to be about speed. Why is it important for my child to learn basic computation facts and compute quickly?

Math shouldn't be associated with speed (Boaler 2014)! But knowing basic facts so that they can be quickly recalled facilitates problem solving and helps children complete math without getting bogged down with toilsome paper-and-pencil computation. Teachers encourage students to compute flexibly, choose an appropriate strategy to use, be efficient, and be accurate (Kling and Bay-Williams 2014). We want children to be able to compute efficiently, perhaps by using the mental math strategies listed in previous questions in this chapter. Having efficient computation strategies will allow your child to get to the meat of mathematics—problem solving—without being mired in paper-and-pencil computation. Children will feel more confident about themselves as doers of math if they have their facts down. Finally, being able to compute quickly allows children to see relationships between numbers and think flexibly.

All this being said, a math-positive parent helps children see that mathematics is much more than being quick with computation. At home, encourage children to practice facts for a few minutes but also reinforce problem solving, reasoning, and real-world connections to math. Share a math problem you encountered at work. Tell how you used logic to problem-solve your way to a solution. Math is much more than adding and subtracting, multiplying and dividing. Math is a way to make sense of the world. For more insights on fluency, see Chapter 2, Question No. 20, *Aren't kids who can do math fast considered better at math than kids who are slower at doing it?*

20. How can I help my child learn the basic facts?

While I do not believe that speed is essential to success in math, being able to make simple calculations efficiently is important to moving on to more complex concepts in math. Thus, teachers may have a system for encouraging children to "pass off" or show mastery of basic computation facts. Encourage your child to spend five to seven minutes several times a week working to build fluency. I hate flash cards but feel justified in allowing my son Duncan to use the family computer or handheld device for focused practice with basic facts. For our family, driving in the car is the best time to practice math facts. Whether you're sitting in traffic or whizzing down the road, you can quiz your child or have them use a math fact app on a device. To find apps that have been evaluated for their effectiveness, functionality, and child-friendly design, search online for reviews and recommendations from organizations like the National Council of Teachers of Mathematics, the National Association for the Education of Young Children, and Children's Technology

Review. One of my favorite apps comes from PBS Kids. The games based on the Cyberchase and Curious George programs are great for building fluency with engaging contexts and well-loved characters.

TEACHING TIP

Thinking about Fact Fluency

Traditional mathematics classrooms have often included timed tests as a means of evaluating fluency with basic facts. Recent research, however, has shown timed tests to be ineffective—even impediments to students' learning, achievement, and attitudes about mathematics. (See research from Jo Boaler [2014] on timed tests. To learn ways to assess fluency with basic facts, see research from Gina Kling and Jennifer M. Bay-Williams [2014].) A math-positive classroom environment includes child-centered assessments of fact fluency and recognizes that children will be ready for fluency practice at different times, and they will progress at different rates. Make strategies explicit in the classroom through number talks that teach and provide practice in using and selecting appropriate, efficient strategies (Parrish 2010, 2014). Have students keep private or individual progress charts and provide intermittent rewards for completing certain sets of facts, if desired. Remember: Students who advance more slowly and who are relegated to a drill of facts when the rest of the class is engaged in meaningful experiences will soon feel left out, lost, and incapable of doing "real" mathematics. These practices firmly plant a fixed mindset in children. By contrast, when students who have not yet mastered facts are engaged in exciting and meaningful experiences, they have real motivation to learn facts and real opportunities to develop relationships that can aid in that endeavor. A math-positive teacher does not allow students who are behind in fact mastery to fall behind in mathematics.

"PAUSE"ATIVE BOX

Tips for Building Math Fact Fluency with a Math-Positive Mindset

Take a moment to pause and review the following tips. Put an asterisk by those tips that you'd like to concentrate on in helping children learn basic facts.

- Realize that children will be ready for fluency practice at different times, and they will progress at different rates.
- Provide practice selecting appropriate, efficient strategies like *Bridging to a 10*, *Using a Known Fact*, and so on.
- Have children create their own personalized set of flash cards with only those facts they need to focus on.
- Present facts rapidly to children—through flash cards, verbally, or using an app or website. Do not give children a page of facts to complete within a set time. Timed tests have been shown to be counterproductive to learning math facts (Boaler 2014).
- Don't stop and explain a fact during a practice session. Explain or ask questions after the session is over.
- Provide games that give fact practice. The Internet is full of drill and practice games. Some of my favorites are Cool Math for Kids, Fun Brain's math zone, and the National Council of Teachers of Mathematics *Illuminations* interactives.
- Use downtime in the car and during waiting periods to practice. Download apps that give quick practice.
- Set goals and provide intermittent rewards for developing fluency with a certain set of facts, if desired.

21. How can I help my child make sense of fractions?

You've probably heard the joke that three out of two people struggle with fractions. Or the one about the inventor of fractions—Henry the Eighth. Fractions are the punch line for a lot of math humor. But keeping fractions real world will do a lot to help children make sense of these useful bits of numbers.

To fracture, of course, means to break. Your child, like a few of my children, might make a personal connection to the concept of a fraction in mathematics by building upon an experience with broken bones.

Fractions can be explored using three main models—area (region), set, and length (sometimes with a number line or strip diagram).

Model 1: Area (Region)

In the area or region model, a whole is partitioned into equal parts, each taking up a fraction of the area of the whole. Circles, rectangles, pattern blocks, and folded paper work well for exploring the area model. A real-world example might involve cutting a square pan of brownies into twelve equal parts to make twelfths.

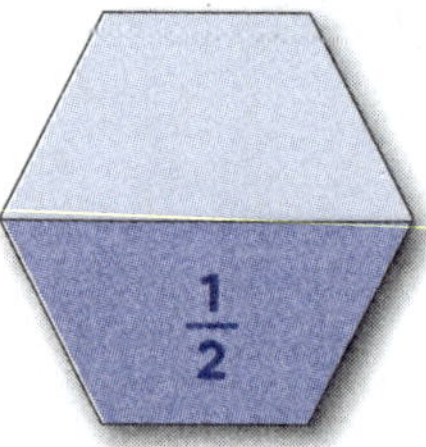

This is an area model for fractions using pattern blocks. The hexagon-shaped pattern block represents the whole. The trapezoid takes $\frac{1}{2}$ of the area (or space) of the hexagon. So the trapezoid is $\frac{1}{2}$ of the hexagon.

Model 2: Set

In the set model, the whole is a set of objects, and subsets of the whole make up the fractional parts. For example, $\frac{7}{8}$ of my children are boys. You can read this as seven out of eight are boys. You can use a set of toy cars to explore the set model with your child. To do this, grab a handful of cars and sort them by, for example, color. If you have ten cars total and three are red, $\frac{3}{10}$ are red.

This photo provides an example of a set model for fractions. The eight Cutler kids you see here (Will, Sybil, Duncan, Chas, Zeb, McGregor, Quinn, and Knox) make the whole set. The fraction of the set that is boys is $\frac{7}{8}$. The fraction that is girls is $\frac{1}{8}$. Poor Sybil. But being the only girl has its perks. My husband, Chris, tells her he divides his love equally— $\frac{1}{2}$ for the boys and $\frac{1}{2}$ for the girls. Is that fair? Probably not, but she is OK with it!

Model 3: Length (Number Line or Strip Diagram)

In the length model, folded paper strips or number lines help children to compare linear measurements instead of areas. Cuisenaire rods and fraction bars (sometimes called bar models) also work well for showing a length model for fractions. The length model is a bit more sophisticated than the set or area model, but it helps children compare the relative sizes of fractions. Try it out by having children fold strips of paper into halves, fourths, and eighths. Then have children label the pieces and cut them apart. Creating these length models draws attention to a big idea in fractions: as the strips get shorter, the denominators (bottom numbers) increase. What do you notice

about the fraction strips? What comparisons can you make? What equivalences can you find?

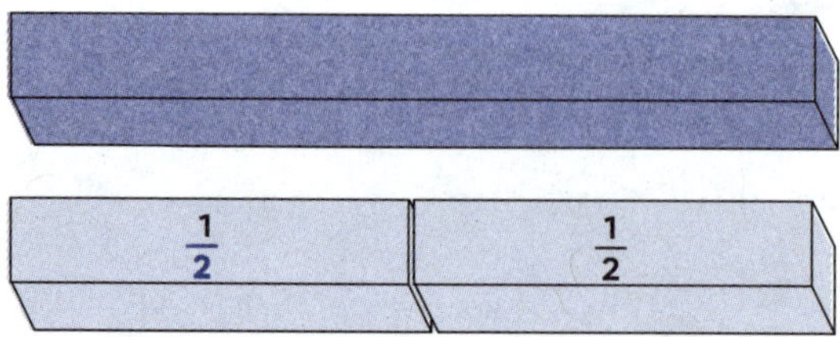

Using Cuisenaire rods, we can show a length model. The top rod represents the whole. The two rods below are each $\frac{1}{2}$ the length of the top rod.

If you're using the length model to perform operations, the partitioning process helps children connect the whole with its composing fractions by connecting to the idea of equal shares. Try this problem to see how a strip diagram works.

The kitchen pantry contains 54 cans. Four-ninths of the cans are fruit and the rest are vegetables. How many cans are fruit?

Try it. Draw a long rectangle for your strip diagram. This is the whole—the total number of cans in the pantry. Partition the strip into nine parts and then figure out the size of each partition. In this case fifty-four divided into nine parts makes equal shares of six. Four groups of six equals twenty-four. There are twenty-four cans of fruit in the pantry. The rest, thirty cans, are vegetables. Yum!

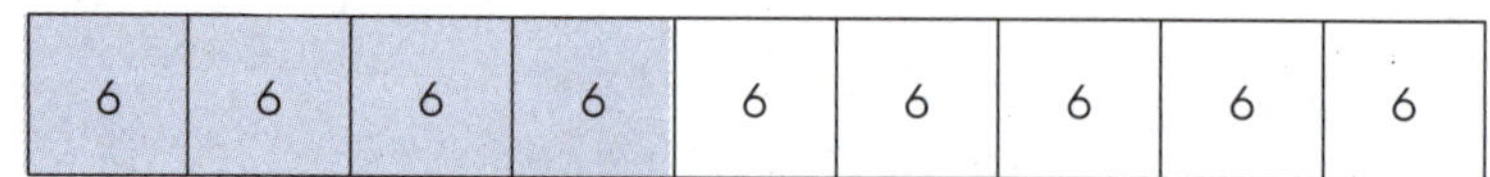

This problem can also be modeled using a number line. Draw the line and label the ends 0 and 54. Partition the line into nine equal parts and label them. Four of the partitions are fruit. That's twenty-four cans.

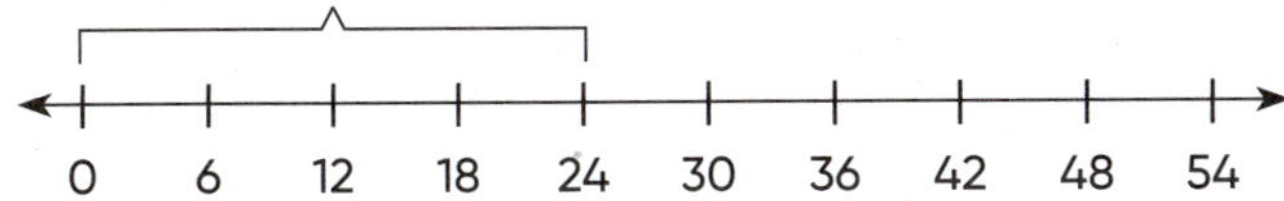

Or you can use bar models, another length model for fractions, to figure out the problem. Here is a drawing of the bar model showing the whole labeled 54 cans. The nine partitions are also each labeled 6. To find $\frac{4}{9}$ of 54, we can add 6 + 6 + 6 + 6 = 24. Cool, huh?

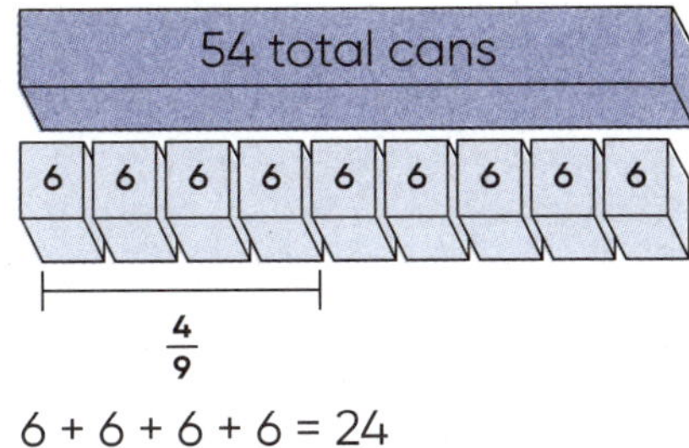

22. I've heard something about benchmark fractions. What are they and how are they helpful?

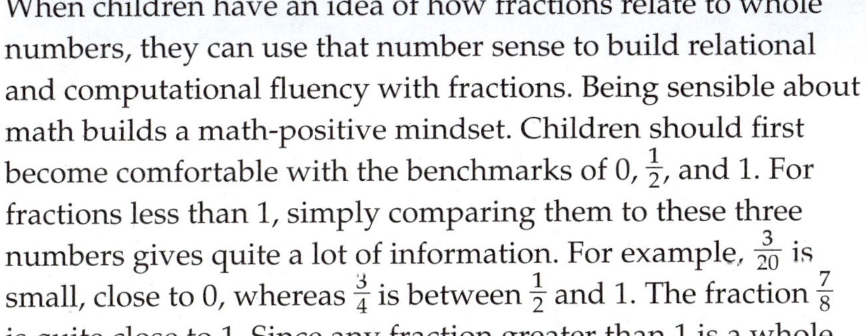

When children have an idea of how fractions relate to whole numbers, they can use that number sense to build relational and computational fluency with fractions. Being sensible about math builds a math-positive mindset. Children should first become comfortable with the benchmarks of 0, $\frac{1}{2}$, and 1. For fractions less than 1, simply comparing them to these three numbers gives quite a lot of information. For example, $\frac{3}{20}$ is small, close to 0, whereas $\frac{3}{4}$ is between $\frac{1}{2}$ and 1. The fraction $\frac{7}{8}$ is quite close to 1. Since any fraction greater than 1 is a whole number plus an amount less than 1, the same reference points are just as helpful: $3\frac{3}{8}$ is almost $3\frac{1}{2}$.

TEACHING TIP

Supporting the Development of Benchmarks for Fractions

Helping students develop benchmarks for fractions is essential to their fraction number sense as it helps them reason about the relative size of fractions. A sorting activity where students classify fractions as Close to 0, Close to $\frac{1}{2}$ or Close to 1 can help students use 0, $\frac{1}{2}$, and 1 as benchmarks. Have students justify their reasoning with a picture or verbal explanation of how they use the numerator and denominator to pick which group a fraction belongs in. Strong number sense, whether for fractions or any other idea in number, contributes to a math-positive mindset, so keep number sense at the forefront of your mind when planning math experiences for your students.

23. What are some tips for helping my child see real-world applications for fractions?

Fractions are everywhere. With children, shop the clearance rack for half-price bargains. Challenge them to figure out the final cost of an item. Next time you cook a recipe that includes fractions of teaspoons, tablespoons, or cups, point this out to children. To make the concept of fractions more concrete, have them completely fill the 1-cup measuring cup using the $\frac{1}{2}$ cup. Dump out the contents and repeat the experiment with the $\frac{1}{3}$ cup, and finally the $\frac{1}{4}$ cup. If your child is learning to play a musical instrument or read music, draw a connection between fractions and musical notations such as quarter notes, half notes, whole notes, and so on.

24. I know the conventional way to write fractions, with a top and a bottom number separated by a bar, but what do these symbols mean?

The symbols are, admittedly, somewhat arbitrary but have been agreed upon as a consistent form for writing fractions. It helps to know that the numerator (the top number) is the counting number. It tells how many equal parts we have. The denominator (the bottom number) tells what or what size piece is being counted. Using all three models for fractions (discussed in Question No. 21) helps children connect the language to the visuals.

25. What are the big ideas for understanding fractions in elementary school?

Just ask any young child if their half of the cookie is fair, and you'll see how adept they are at judging equal portions, equality, and fair shares. Formal investigation of fractions generally begins in third grade where children work with fractions with denominators 2, 3, 4, 6, and 8. In this foundational phase, they work hard to develop number sense for the size of different fractions by using many models including folded paper, fractions bars, and number lines. Children also begin to explore equivalence among fractions. In fourth grade, children build on the previous year's learning by relating fractions with the denominator 10 and 100 to decimals. Fourth graders also begin to explore operations with fractions. Students who can generate equivalent fractions can develop strategies for adding and subtracting fractions with unlike denominators, but addition and subtraction with unlike denominators is not an emphasis until grade 5. In grade 5, children also divide fractions by whole numbers and whole numbers by fractions. This is where the region models, bar models and strip diagrams discussed in Question No. 21 really come in handy.

"PAUSE"ATIVE BOX

Summary of Big Ideas for Fractions for Grade 3 through Grade 5

Here are some of the big ideas for fractions children will learn in school, including activities you can do to support their learning. The ideas are connected to the Common Core State Standards (National Governors Association Center for Best Practices and Council of Chief State School Officers 2010). Pause and take a moment to review the list. Put an asterisk next to some of the activities you'd like to try. Try them with your child. How did it go?

<table>
<tr><th>Grade</th><th>Big Ideas in Fractions</th><th>Math Activities</th></tr>
<tr><td rowspan="5">3</td><td>Understand fractions as parts of a whole.</td><td>The best way to introduce fractions is with stories. Focus on contexts that are easy-to-draw and partition. For a set model, try stories about sharing grapes equally with friends. For example, Three children want to fairly share 15 grapes. How many grapes will each child get? For an area model, talk about sharing 5 pancakes with 4 people. For a length model, try stories about breaking 3 sticks of gum to share equally among 2 people. There are no "fractions" in these stories; the partitioning (sharing) comes as children draw.</td></tr>
<tr><td>Represent a fraction using a number line.</td><td>Have children create a number line by folding a strip of paper in half. Work together to label 0, $\frac{1}{2}$, and 1. Fold in half again and label the locations for $\frac{1}{4}$ and $\frac{3}{4}$. Talk about how the strip of paper looks like a number line.</td></tr>
<tr><td>Understand two fractions as equivalent if they are the same size or the same point on a number line.</td><td rowspan="3">To build understanding of equivalence and comparisons with fractions, create a Fraction Bar Kit (Stenmark, Thompson, and Cossey 1986). You will need 5 strips of paper about 2 inches wide and 8 inches long.
• Whole: Keep one strip whole and label it "1 whole."
• $\frac{1}{2}$: Take another strip and fold it carefully in half widthwise. Have children guess how many sections it will have when unfolded. Label each part $\frac{1}{2}$ and cut on the folded line.
• $\frac{1}{4}$: Take another strip and fold it carefully in half widthwise 2 times. Have children guess how many sections it will have when unfolded. Count the sections, label each part $\frac{1}{4}$ and cut them apart.
• $\frac{1}{8}$: Take another strip and fold it carefully in half 3 times. Have children guess how many sections it will have when unfolded. Count the sections, label each part $\frac{1}{8}$, and cut them apart.</td></tr>
<tr><td>Recognize and generate simple equivalent fractions (e.g., $\frac{1}{2} = \frac{2}{4}$). Use models to explain why the fractions are equivalent.</td></tr>
<tr><td>Express whole numbers as fractions (e.g., $6 = \frac{6}{1}$) and recognize fractions that are equivalent to whole numbers.</td></tr>
</table>

(continued)

(continued)

Grade	Big Ideas in Fractions	Math Activities
3	Compare two fractions with the same numerator or the same denominator by reasoning about their size. Use >, =, or < symbols and justify the conclusions using a visual fraction model.	• $\frac{1}{16}$: Take another strip and ask children how many times they think they should fold it. Fold in half 4 times. Guess how many sections, unfold, count, label each part $\frac{1}{16}$, and cut the parts apart. Explore with the pieces. Find combinations that make a whole. Notice how the smaller pieces have a "big" number for the denominator. Explain that the denominator (the bottom number) tells what size piece is being counted. The larger the number, the more pieces and the smaller they have to be in order to fit on the whole strip.
4	Use models to explain why fractions with different numbers and sizes of parts are equivalent. Recognize and generate equivalent fractions.	Use the Fraction Bar Kit described in grade 3 of this table to reinforce equivalence. Consider adding thirds, sixths, and twelfths to the kit to discover even more equivalent fractions by laying the pieces on top of one another.
	Compare two decimals to hundredths by reasoning about their size. Use >, =, or < symbols and justify the conclusions using a visual model.	Decimals are easily compared when drawn on two 10×10 grids. The 100 boxes represent a whole and colored-in spaces are the parts of the whole—decimals. For example, to compare 0.75 and 0.57, color in 75 boxes on the first grid and 57 boxes on the second grid. Remember to say the decimals numbers correctly to draw a connection to fractions. 0.75 is said "75 hundredths" and 0.57 is said "57 hundredths."
	Compare two fractions with different numerators and different denominators by creating common denominators or numerators or by comparing to a benchmark fraction such as $\frac{1}{2}$. Use >, =, or < symbols and justify the conclusions using a visual fraction model.	Comparing decimals with different denominators works best with a physical model. Try using fraction bars or the Fraction Bar Kit described in grade 3 of this table to explore relationships between fractional pieces.
	Understand addition and subtraction of fractions as joining and separating parts that come from the same whole.	The Fraction Bar Kit described in the grade 3 section of this table is perfect for developing these concepts. Give children a lot of hands-on experience with models before moving on to the abstract symbols. Here is a fun game to try: *Cover Up* (Burns 1984) Materials: Fraction Bar Kit (with whole, $\frac{1}{2}$, $\frac{1}{4}$, $\frac{1}{8}$ and $\frac{1}{16}$ pieces), die marked with these sides: $\frac{1}{2}$, $\frac{1}{4}$, $\frac{1}{8}$, $\frac{1}{8}$, $\frac{1}{16}$, $\frac{1}{16}$
	Decompose fractions and use models to justify decompositions (e.g., $\frac{3}{4}$ can be thought of as $\frac{1}{2} + \frac{1}{4}$ or $\frac{1}{4} + \frac{1}{4} + \frac{1}{4}$).	

(continued)

(continued)

Grade	Big Ideas in Fractions	Math Activities
4	Add and subtract mixed numbers with like denominators. Use equivalent fractions, properties of operations and/or the relationship between addition and subtraction.	Instructions: 1. Each player works with their own Fraction Bar Kit. The object of the game is to be the first player to completely cover the "whole" strip with the other fractional pieces from their kit without overlapping. 2. Take turns rolling the die and covering the "whole" strip with the fractional pieces. 3. If you roll and get a fraction larger than there is space on your "whole" strip, you lose a turn and must wait until you roll a fraction that will fit. For example, if you only have $\frac{1}{8}$ space remaining on your "whole" strip, rolling a $\frac{1}{2}$ cannot be used. 4. The first person to cover up their entire "whole" strip wins. For an added challenge, play *Uncover* (Burns 1984). 1. Use the same die and fraction pieces but start with the 2 halves covering the "whole" strip. 2. Roll and remove the fraction shown on the die. 3. An exchange or trade may be necessary before the fraction can be removed. Use your knowledge of equivalence to help you. For instance, if you roll $\frac{1}{8}$ on the first roll, one of the $\frac{1}{2}$ pieces needs to be exchanged for $\frac{4}{8}$ in order to remove $\frac{1}{8}$. Each player must approve of the exchanges. Watch carefully and talk through your reasoning. 4. The first person to uncover their entire "whole" strip wins.
	Solve word problems involving addition and subtraction of fractions.	
	Understand a fraction such as $\frac{5}{8}$ as a multiple of $\frac{1}{8}$, so $5 \times \frac{1}{8} = \frac{5}{8}$.	Introduce this concept by using kid-friendly stories. Here is one to talk about and solve together by drawing pictures. *Amir is baking cookies. He needs $\frac{2}{3}$ of a cup of chocolate chips for each batch of cookies. How many cups of chocolate chips does he need for 5 batches of cookies?* Talk about what is meant by 5×2 (5 groups of 2) and what is meant by $5 \times \frac{2}{3}$ (5 groups of $\frac{2}{3}$). Your picture might look something like this: Children can then count to find how many chocolate chips are needed: $\frac{10}{3}$ cups.
	Solve word problems involving multiplication of a fraction by a whole number.	

(continued)

(continued)

<table>
<tr><th>Grade</th><th>Big Ideas in Fractions</th><th>Math Activities</th></tr>
<tr><td rowspan="2">4</td><td>Express a fraction with denominator 10 as an equivalent fraction with denominator 100, and use this technique to add two fractions with respective denominators 10 and 100.</td><td rowspan="2"> Choral Counting (Franke, Kazemi, and Turrou 2018) with decimal fractions helps build the idea that decimals are simply a different representation of fractions. Use equivalent fractions in choral counts to explore how fractions with different denominators can represent the same value. A choral count by tenths could be followed by a choral count by hundredths, for example. (Read about choral counting under Question No. 7 of this chapter.) Tell children to count aloud, adding $\frac{1}{10}$ each time. Put a twist on traditional choral counting by having children close their eyes as they count. Record their counts as decimals instead of fractions. When children open their eyes, they may be surprised to see decimals when they envisioned a fraction representation. For a fun extension, record some of the numbers as fractions and others as decimals.
<table>
<tr><td>$\frac{1}{10}$</td><td>0.2</td><td>$\frac{3}{10}$</td><td>0.4</td><td>$\frac{5}{10}$</td></tr>
<tr><td>0.6</td><td>$\frac{7}{10}$</td><td>0.8</td><td>$\frac{9}{10}$</td><td>1</td></tr>
</table>
To compare sizes of decimals, have children color in the decimal as represented on a 10 × 10 grid. It's easy to see how a decimal is larger if it covers more space. This is trickier to "see" on a number line. Try this activity to draw that connection: have children cut the colored boxes of the grid into strips and tape them into a single long piece. Do this for the comparison decimal as well. Which strip is longer? By how much? Children can count the boxes to find out. If you want to, you can cut apart a blank 10 × 10 grid and make this a number line. Lay the other strips below the whole number line and see how far they extend. </td></tr>
<tr><td>Use decimal notation for fractions with denominators 10 or 100. For example, rewrite 0.55 as $\frac{55}{100}$, describe a length as 0.55 centimeters, and locate 0.55 on a number line.</td></tr>
<tr><td>5</td><td> Add and subtract fractions with unlike denominators by using equivalent fractions (e.g., $\frac{2}{5} + \frac{5}{3} = \frac{6}{15} + \frac{25}{15} = \frac{31}{15}$). Solve word problems involving addition and subtraction of fractions including cases of unlike denominators. Use benchmark fractions and number sense to estimate and assess the reasonableness of answers. </td><td> Rather than rush to a procedure for finding common denominators, give children a lot of experience building common fractions. Here is a problem to get you started: Nayeli did $\frac{1}{2}$ of her homework in the morning. She did $\frac{2}{5}$ of her homework in the afternoon. How much homework has Nayeli finished? Using bar models, set the $\frac{1}{2}$ piece and two $\frac{1}{5}$ pieces beside each other. To find what fractions these pieces have in common, look for other fractional pieces (denominators) that fit both fractions. For this problem, the $\frac{1}{10}$ pieces can be used to find equivalences for $\frac{1}{2}$ and for $\frac{1}{5}$. It takes $\frac{9}{10}$ to equal the same length as $\frac{1}{2} + \frac{2}{5}$. (See next page.) </td></tr>
</table>

(continued)

(continued)

<table>
<tr><th>Grade</th><th>Big Ideas in Fractions</th><th>Math Activities</th></tr>
<tr><td rowspan="4">5</td><td></td><td> 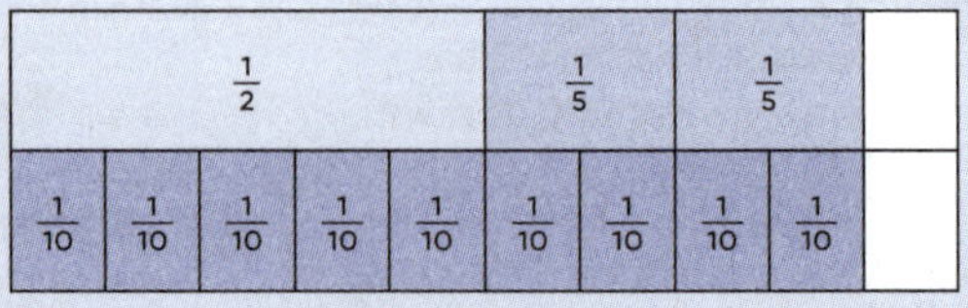$\frac{1}{2}$ $\frac{1}{5}$ $\frac{1}{5}$
$\frac{1}{10}$ $\frac{1}{10}$ $\frac{1}{10}$ $\frac{1}{10}$ $\frac{1}{10}$ $\frac{1}{10}$ $\frac{1}{10}$ $\frac{1}{10}$ $\frac{1}{10}$
Try the same modeling process for a subtraction problem:
Marta has $\frac{3}{4}$ of a candy bar. She eats $\frac{1}{2}$ of her candy bar. How much does she have now?</td></tr>
<tr><td>Interpret a fraction as division of the numerator by the denominator, especially in word problems such as: If 12 dogs must share a 15-pound bag of dog food equally by weight, how many pounds of food should each dog get? Between what two whole numbers does your answer fall on the number line?</td><td rowspan="2">Victor had $\frac{3}{4}$ of a carton of eggs. He used $\frac{1}{6}$ of the eggs to make cookies. What part of the carton of eggs did he use?
Draw a picture to help. First show $\frac{3}{4}$ by coloring in $\frac{3}{4}$ of the rectangle. Notice that the lines are drawn vertically to partition the rectangle. Then partition the same box into sixths using horizontal lines. Color in $\frac{1}{6}$. (This is shown in the second part below, but just draw one box.) What is $\frac{1}{6}$ of $\frac{3}{4}$? The overlapping coloring shows the numerator—3. Count how many little squares are in the whole box to find the denominator—24.
$\frac{1}{6} \times \frac{3}{4} = \frac{3}{24}$ 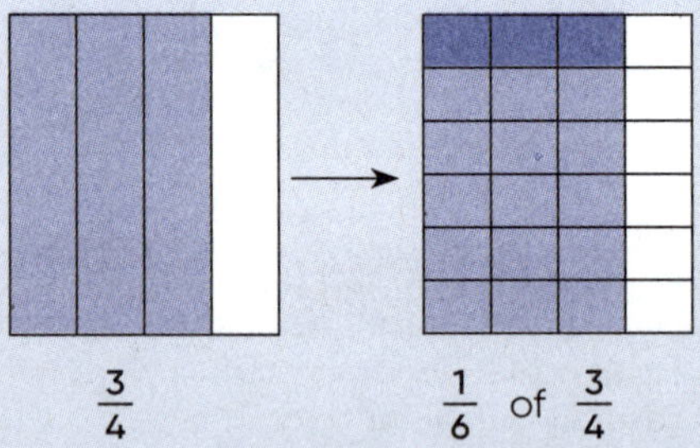
$\frac{3}{4}$ → $\frac{1}{6}$ of $\frac{3}{4}$
Encourage children to talk aloud as they draw the rectangles and partition them. The focus should be on making sense of why this works rather than doing it fast or automatically.</td></tr>
<tr><td>Interpret the product of a fraction multiplied by a whole number as parts of a partition. For instance, a 6-foot-by-1-foot roll of paper cut into $\frac{1}{2}$-foot strips. Use a visual fraction model to show $\frac{1}{2} \times 6 = \frac{6}{2}$ and create a story context for this equation.</td></tr>
<tr><td>Find the area of a rectangle with fractional side lengths by tiling it with unit squares and by multiplying the side lengths to show the same area.</td><td>Area models work well for this concept. Draw the connection to area using whole numbers with an emphasis on the meaning of area—the amount of space inside a figure. Talk with children about how these two problems are similar but unique:</td></tr>
</table>

(continued)

(continued)

Grade	Big Ideas in Fractions	Math Activities
5		Problem A A country garden measures 10 feet by 7 feet. What is its area? Problem B A plot in an urban neighborhood garden measures $\frac{3}{4}$ feet by 7 feet. What is its area?
	Interpret multiplication as scaling (resizing) by comparing the product to the size of one factor.	Scaled drawings can be fun ways to integrate art with math. Try this with children: each person draws a simple design with straight-ish lines (pine trees, sail boats, and flags work well) on a piece of centimeter grid paper. Trade papers and "blow up" the picture by making all the lines 2 times longer. For example, if drawing a pine tree, the trunk on the original might be 6 centimeters long and 2 centimeters wide, but on the scale drawing the trunk is drawn 12 centimeters long and 4 centimeters wide. Children might have to tape together several sheets of centimeter paper to have enough space for their double-sized drawing. Now make a scale drawing that is $\frac{1}{2}$ as big as the original. This time the pine tree's trunk is 3 centimeters long and one centimeter wide. Display all three masterpieces together and show off how multiplication by a whole number or a fraction resizes the drawings. This is an art and math project worthy of hanging on the fridge!
	Explain why multiplying a number by a fraction greater than 1 results in a product greater than the number (e.g., why does $\frac{3}{2} \times 8$ result in a product larger than 8?) and why multiplying a number by a fraction less than 1 results in a product smaller than the number (e.g., why does $\frac{1}{2} \times 8$ result in a product smaller than 8?)	Talk about these problems together to make sense of the fractions hiding inside (composing) whole numbers: *How many quarter-hours are in five hours?* *How many thirds of a cup are in four cups?* *How many half-meters are in six meters?* *How many eighths of a mile are in three miles?* If it helps to draw a picture to talk through the scenarios, do it! The answers are greater than the whole number in each scenario because they have more than 1 group of the fraction.
	Relate equivalence to the effect of multiplying a fraction by 1 (e.g., $\frac{3}{4} \times \frac{6}{6} = \frac{3}{4} \times \frac{8}{8}$).	When children recognize that "wholes" can be represented using an endless number of fractions, they can use different wholes flexibly for solving problems. Which "whole" fraction would work well for the following problem? *Tanner had $\frac{3}{4}$ of a case of soda leftover after a party. If the case had 24 cans, how many sodas are left?*
	Solve real-world problems involving multiplication of fractions and mixed numbers.	

(continued)

(continued)

Grade	Big Ideas in Fractions	Math Activities
5	Create a story context and use models to find $\frac{1}{8} \div 3$.	Division of fractions doesn't have to be relegated to rote procedures like "invert and multiply."
	Use the relationship between multiplication and division to explain that $\frac{1}{8} \div 3 = \frac{1}{24}$ because $\frac{1}{24} \times 3 = \frac{1}{8}$.	Have children draw and talk while they visualize what's going on in this division scenario: *I have three large chocolate candy bars that are perforated into eight pieces each. If I divide the bars into these sections how many pieces will I have?*
	Solve real-world problems involving division of fractions by whole numbers and division of whole numbers by fractions.	Then introduce the problem element to the scenario: *If I divide the bars into these sections what will be the size of each piece?* Drawing a picture helps as does talking it through. Here is a problem for $\frac{1}{4} \div 3$: *K'Mera has one-fourth of a pan of brownies leftover from a party. She wants to share the brownies with her three friends. What fraction of the whole pan of brownies will each friend get?* Draw the whole pan. Use vertical lines to partition into fourths and make a bold border around $\frac{1}{4}$. Then partition the whole pan into three pieces using horizontal lines. How many pieces are there in the pan? There are 12 pieces. If three people each get a share, how big are their shares? They are $\frac{1}{12}$ of the original pan of brownies—bite-sized!

26. What are some materials to help children understand number concepts?

Children's classrooms are hopefully stocked with materials to help them gain hands-on experience with math concepts. These materials might include cubes and blocks of various sizes, shapes (pattern blocks), proportions, scales, calculators, and so on. Concrete models can help children represent their thinking about a mathematical concept. By "concrete models" we mean something that exists physically in the world and that the student can manipulate or move around. Young children should be active learners of mathematics. Hands-on experiences build math-positive mindsets and provide children with opportunities to explore and solidify their thinking about mathematical ideas. Students gain a deeper understanding of mathematics through connecting their ideas and finding ways to express them. As children get older, we hope they develop the ability to think more abstractly. But for budding mathematicians, concrete materials are essential for developing concepts. At home, you can substitute dry beans, blocks, or any small objects to help your child count, perform operations, measure, and so on. You do not have to spend a lot of money on commercial mathematics manipulatives. However, manipulatives might also make a fun birthday present. My kids can pretty much expect all of their holiday gifts to be educational. In my opinion, you can never have too many blocks or books!

"PAUSE"ATIVE BOX

Manipulatives for Building Number Concepts

Following is a list of some of my favorite manipulatives for building number concepts, including what each manipulative can be used for and activity ideas. Take a moment to pause and read through this list. Put an asterisk by the activities you'd like to try, then try one with your child or students. How did it go?

Manipulative	Description	Uses in Number and Operations	Math Activities
Unifix Cubes and Snap Cubes	Unifix Cubes are colorful cubes that connect in only one way. Snap Cubes (also called Linking Cubes) are similar to Unifix Cubes but they can be connected on all sides (six ways).	• Counting and skip-counting • Sorting and patterning • Addition, subtraction, multiplication, division • Number bonds • Place value • Fractions • Measurement	Make a stick 10 cubes long using just one color. This is a 10 train. Have children find other 10 trains for the number bonds for 10 using two different colors (e.g., 3 red and 7 yellow). Have children hold their train next to the 10 train to make sure it is the same length. Repeat until all the number bonds for 10 are represented. What happens when you turn a train over? The commutative property!

(continued)

(continued)

Manipulative	Description	Uses in Number and Operations	Math Activities
Counters	Counters come in a variety of styles including bears, sea creatures, frogs, dinosaurs, vehicles, farm animals, and many more.	• Counting and skip-counting • Sorting and patterning • Addition, subtraction, multiplication, division • Word problems • Fractions • Measurement	Write the numbers 0–20 in the bottom of paper muffin cups. Have children fill each cup with the correct number of counters. This is great practice for one-to-one correspondence, counting, and recognizing the digits. Don't forget to include 0!
Five-Frames and Ten-Frames	A rectangle with 1 × 5 grid (five-frame) or 2 × 5 grid (ten-frame).	• Counting and counting on • Number composition and decomposition • Bridging to a 10 • Place value • Addition and subtraction	Practice the teen numbers by drawing two ten-frames. Put 10 dots in the first ten-frame. Fill the second ten-frame with dots and have children begin counting with 10 or tell the total dots on both ten-frames without counting at all. Teen numbers can be thought of as "ten and some more."

Base Ten Blocks	Plastic blocks consisting of:
	• units (one cube),
	• longs (10 units),
	• flats (10 longs or 100 units), and
	• cubes (10 flats or 1,000 units).
		• Place-value concepts in our base ten system:
		  • Zero when used as a place holder
		  • Regrouping
		  • Addition, subtraction, multiplication, and division
		  • Fraction concepts
		  • Connecting written numbers with their values
		  • Number composition and decomposition
		  • Expanded notation
			Explore with blocks to practice regrouping.
			Try 23 + 18.
			Build 23. Build 18. Join the sets. Count. What can you trade to make counting easier? Trade 10 ones for a rod (one 10). Then skip count the 10s and add the 1 at the end.

(continued)

(continued)

Manipulative	Description	Uses in Number and Operations	Math Activities
Hundreds Charts	A 10 × 10 grid numbered from 1 to 100.	• Building visual understanding of patterns in our base ten system • Counting and skip-counting • Number magnitude comparisons • Addition and subtraction	Try these fun challenges: • Cut apart the chart and have children put it back together. • Have children color all the even numbers and tell about the patterns they see. • Have children tell what number is 10 more, 10 less, 1 more, and 1 less than a target number.
Two-Color Counters	Chips with a different color on each side, usually red and white or red and yellow.	• Addition, subtraction, multiplication, division • Fractions of a set	Use the counters to explore the set model for fractions. Have children put 8 counters in a cup and spill them on the table. Count how many are red and how many are white. Use fractions to describe the parts of the set. For example, $\frac{3}{8}$ (3 out of 8) chips are red. Another example is $\frac{5}{8}$ (5 out of 8) chips are white.
Cuisenaire Rods	Colored rods that are proportional to one another. The set consists of rods in the following sizes and colors: • 1 cm. (white), • 2 cm. (red), • 3 cm. (light green), • 4 cm. (lavender), • 5 cm. (yellow), • 6 cm. (dark green), • 7 cm. (black), • 8 cm. (brown), • 9 cm. (blue), and • 10 cm. (orange).	• Measuring length • Exploring ratios and proportions • Number composition and decomposition • Counting • Fractions • Addition and subtraction	Have children explore the rods to find all the ways to make $\frac{1}{2}$. For example, if you think of the orange rod as the whole, then the yellow rods are $\frac{1}{2}$. But if you think of the brown rod as the whole, what would the half be?

(continued)

(continued)

Manipulative	Description	Uses in Number and Operations	Math Activities
Geoboards	A plastic board with pegs around which the child wraps rubber bands to create shapes.	• Primarily used to help build geometry concepts • Fractions	Use geoboards to partition shapes into fractions. For example, have children make a square, then show two ways to divide it in $\frac{1}{2}$.
Fraction Bars and Fraction Circles	Each type consists of proportional colored tiles. A set usually includes: • one red *whole*, • two pink *halves*, • three orange *thirds*, • four yellow *fourths*, • five green *fifths*, • six teal (blue-green) *sixths*, • eight blue *eighths*, • ten purple *tenths*, and • twelve black *twelfths*.	• Visualizing and comparing fractions • Creating equivalent fractions • Addition and subtraction of fractions	When it comes to fractions, many people think of pizza—a handy real-world reference for making sense of partitioning. Ask these questions to start a conversation about fractions with children: • Would you rather eat 1 slice of a pizza that is cut into fourths or sixths? How can you use the fraction circles to show your thinking? • Would you rather eat 2 slices of a pizza that is cut into thirds or 4 slices of a pizza that is cut into sixths? Why is this a trick question?

Pattern Blocks	Plastic or wooden blocks that come in six shapes:
	• yellow hexagons,
	• orange squares,
	• red trapezoids,
	• green triangles,
	• tan rhombi, and
	• blue rhombi.
		• Classification and patterning
		  • Area models for fractions (for example, a green triangle is $\frac{1}{6}$ of a yellow hexagon)
		  • Set models for fractions (for example, four out of six pattern blocks are quadrilaterals)
		  • Number patterns that repeat, shrink, or grow
			Make a fraction book. Fold three sheets of paper in half and staple. On each page, have children trace the yellow hexagon. This will be our whole. Have children find other pattern blocks that, when combined, make a whole. They can lay the pattern blocks on top of the hexagon to see if the pieces compose 1 whole hexagon. For an added challenge, label the pieces with their fraction name. For example, the green triangles are each worth $\frac{1}{6}$ of a hexagon.

(continued)

(continued)

Manipulative	Description	Uses in Number and Operations	Math Activities
Random Number Generators	These tools provide numbers by chance and can be used in games or activities. Examples of these tools include dice, spinners, decks of playing cards or numbered cards, and dominoes.	• Practicing spatial subitizing • Operations (i.e., multiply the numbers shown) • Comparing numbers • Finding number bonds	Remove the face cards from a deck of playing cards. Play *Sums of Ten Go Fish*. Each player gets seven cards. The rest of the cards are the fishing pond. They look in their hand for two cards with a sum of 10 (number bonds for 10). If they find a number bond, they set it aside. Take turns asking other players for the cards needed to complete a number bond. For example, if you have a 5 in your hand, ask another player for a 5. If they don't have a 5, you "Go Fish" by picking up a card from the fishing pond. The winner is the player with the most number bonds for 10 when all the cards have been used.

QUICK REFERENCE CHART

Common Concerns about Number and Operations

This table is a quick reference for some of the *most common concerns* I hear from parents. The table lists the concern, an explanation of the thinking behind it, and an idea that addresses the concern. Consider photocopying this table and placing it somewhere where you can be reminded of what is needed to build your math-positive mindset and that of your children and/or students.

Concern	Explanation	What to Do	Question Number(s) (to Learn More)
Counting difficulties	Competent counters must know the number sequence, match a number to each item counted, keep track of what's been counted, and recognize that the final number said is the total number in the set.	Practice meaningful counting while washing hands, picking up toys, or climbing the stairs. With practice, children will move on to more sophisticated types of counting, like beginning from a number other than one—a skill that leads to counting on from the larger number when adding.	1, 2
What is subitizing?	The ability to know the number of items in a small set (usually up to five items) without counting.	Play dice games to practice conceptual subitizing by quickly recognizing the number of pips on a die instead of counting them individually.	3
My child skips numbers.	A child who has not yet developed one-to-one correspondence may say the number sequence correctly when counting aloud by rote but when counting items, skips numbers. Don't worry, children will get it in time.	Help children develop one-to-one correspondence by having them pick up and move items to the other side of the table, counting each as they do so. Gently put your hand on children's hands and help them point and count slowly. Then have each child do it again by themselves, touching each item as they count aloud.	4
My child writes numbers backward.	Writing the numbers requires coordination of the hand, fingers, and brain. Writing digits backward, especially 2 and 5, is common among young children and a normal part of development.	Post problematic digits such as 2 and 5 or 9 and 6 in prominent places around your home or classroom. You might label your child's dresser drawers so that they see these digits more often. Draw attention to how the digit is formed by placing a dot at the spot where they would first place their pencil.	5

(continued)

(continued)

Concern	Explanation	What to Do	Question Number(s) (to Learn More)
What are number bonds?	The two addends that make a whole number is a number bond. For example, the number 7 can be composed of 6 and 1, 5 and 2, 3 and 4, and 0 and 7. Knowing number bonds helps children to think flexibly about numbers—to pull them apart and put them back together again in useful ways.	Practice number bonds with dominoes. Tell children the total number of pips but cover one side of the domino with your finger. Children tell you what's covered. Ask children to tell you another way to split the total number of pips.	6
What is number sense?	Encompasses estimation, number magnitude and relative position, place value, and counting strategies Strong number sense is a lifelong endeavor with real-world applications.	Build number sense by estimating during a shopping trip. How much will a melon weigh? How long will it take to check out? How much will the groceries cost? Sketch a number line on paper. Label *0* and *100* on the ends. Mark your secret number with a ? (estimate as best you can). Children guess. Place a mark and label the guesses until the correct number is guessed.	7, 8
Why the base ten system?	Two general rules govern our base ten system. First, a digit receives its value based on its position in a number. Second, we implicitly trade 10 ones for 1 ten, 0 tens for one hundred, and so on.	Explore our base ten system using base ten blocks. These manipulatives come in units, rods of ten, and flats of one hundred. Build numbers first, then use the blocks for adding and subtracting.	9
Why a hundreds chart?	Supports algebraic reasoning and recognizing patterns in base ten system. Helps build understanding of number composition and decomposition (groups of tens and ones).	Play *Hundreds Chart Puzzles* (cut apart chart and reassemble) and *Hundreds Chart Hunt* (leave off some numbers and have children fill in).	9
Regrouping instead of borrowing	Regrouping language shows how we model the problem not by borrowing something but by changing the way numbers are grouped.	The best way to learn regrouping is not with the standard algorithm (carrying and borrowing) but using objects that require a child to explicitly trade. Base ten blocks (real or virtual) help children see what's really happening when we subtract or add across the ones, tens, and hundreds places.	10

(continued)

(continued)

Concern	Explanation	What to Do	Question Number(s) (to Learn More)
Regrouping instead of borrowing *(cont'd)*		Ask children to show you how to use virtual base ten blocks. Try these: http://www.abcya.com/baseten.htm and www.mathlearningcenter.org.	10
Estimation and the word *about*	Estimation is a real-world mathematics skill. In word problems, the word *about* is sometimes used to indicate that the answer will be an estimate rather than a precise number. *About* means "very close to" or "almost the same as" (Monroe 2006).	Call children's attention to all the moments you use estimation during the day (for example, how much time needed to get ready for school and work, how much money needed to buy something, etc.).	11
Mental math	Mental math requires solving a computation problem without using paper and pencil, a calculator, or manipulatives.	Strengthen mental math by doing a few addition problems each day. Emphasize the strategies found in this chapter like splitting a number or compensation. Agility with arithmetic is a foundation for all other areas of mental math.	12
Addition facts	Addition facts are foundational to the other operations, place value, and number sense. Having a range of strategies rather than relying on memorization supports fluency by allowing children to select from appropriate, efficient strategies for computing.	Play *Addition Top-It* (Bay-Williams and Kling 2014) to build fluency with the addition facts. You'll need a set of number cards with four cards each of the numbers 0–10. Lay the cards facedown on the table. Play with a partner. At the same time, each player turns over two cards and calls out their sum. The player with the highest sum wins the round and gets to keep all four cards. In the case of a tie, each player turns over two more cards and calls out their sum. They player with the highest sum takes all eight cards. The game ends when there are not enough cards left for each player to turn over two cards. The player with the most cards wins.	13

(continued)

(continued)

Concern	Explanation	What to Do	Question Number(s) (to Learn More)
Multiplication facts	Give children ample time to develop the concept of multiplication with lots of hands-on experience using manipulatives. Then build number sense by reasoning about how multiplication facts relate to one another and cluster together (like 6×4 can also be thought of as 6×2 plus 6×2).	Teaching children the zero, identity, and commutative properties of multiplication eliminates the need for learning a bunch of facts. Focus on multiplying by 2, 5, 10 next. Then move on to the multiplication squares (6×6, for example) before tackling the rest of the facts. Learning the facts in this order helps build on known ideas about our base ten system and connects to counting and addition.	15, 16
Subtraction facts and division facts	Subtraction and division facts receive less emphasis than addition and multiplication facts, partly because we hope children see how these operations are the inverse of addition and multiplication.	Instead of drilling subtraction facts, use number bonds explorations to find the missing addend. Use the "think multiplication" strategy as an approach to solving division problems.	14, 17, 18
Fact fluency	Not being slowed by computation helps children move on to meaty, rich problems.	Practice 5–7 minutes several times a week. Do it in the car. Create flash cards. Use apps. Make it fun!	19, 20
Models for fractions	Example of area model: a rectangle is divided into 4 sections and 3 sections are shaded. Example of set model: 3 out of 4 buttons are black and 1 is white. Example of length model: it takes three $\frac{1}{3}$ pieces to equal the length of 1 whole.	Different problems prompt different models for making sense of fractions. Play with the set model by using two colors of breakfast cereal. Talk about it using fractional terms, "Six-eighths of the cereals are colored red. Two-eighths are blue."	21
Benchmark fractions	Comparing unfamiliar fractions to the benchmark fractions of 0, $\frac{1}{2}$, and 1 helps us judge relative size.	Do a sorting activity where children place unfamiliar fractions into piles of Close to 0, Close to $\frac{1}{2}$, or Close to 1. They can justify their reasoning with a picture or verbal explanation of how they use the numerator and denominator to pick which group a fraction belongs in.	22

(continued)

(continued)

Concern	Explanation	What to Do	Question Number(s) (to Learn More)
Where can I find real-world fractions?	Fractions are everywhere—recipes, discounts, gas tank gauges, and so on.	Keep fractions meaningful by pointing out how useful these bits of numbers can be while cooking, shopping, and cruising down the road. Challenge your child to find math in their world!	23
What are numerators and denominators?	The numerator (the top number) is the counting number. It tells how many equal parts there are. The denominator (the bottom number) tells what size each part is.	Memorizing these terms isn't as important as knowing what they mean. Help children make sense of the parts of a fractional representation by using models such as paper strips and drawings.	24
What are fraction concepts for grades 3–5?	Fraction concepts start with representing fractions and discovering equivalent fractions before comparing them. Later, children are ready to add, subtract, multiply, and divide fractions.	Fraction learning should be done with the goal of understanding part-to-whole relationships. Use concrete materials like folded paper strips and drawings to build number sense and conceptual understanding.	25
What materials do I need for building number concepts?	Unifix cubes, counters, five- and ten-frames, base ten blocks, hundreds charts, two-color counters, Cuisenaire rods, geoboards, fraction bars, pattern blocks, and dice all come in handy for building understanding of number.	Math tools make great birthday and holiday gifts. Put them on the shelf next to toys to encourage children to use them often. Children who have frequent exposure to math tools find counting, operations, estimating, and fractions an enjoyable part of their daily math-positive play.	26

I Can Help with Geometry

CHAPTER 5

TEACHING TIPS

1. I thought geometry was a high school class. Why is geometry emphasized so young?

Knowledge of geometry comes into play in fields as diverse as fashion design, computer animation, construction, interior design, plumbing, art, and the development of global positioning systems. And while we wouldn't expect a young child to state and apply the Pythagorean theorem and its converse, we would expect a youngster to be able to sort triangles according to their attributes, an underpinning essential to that more-sophisticated skill.

All learners can develop the ability to think and reason in geometric ways, but this ability requires children to progress through increasing levels of geometric thought. Two Dutch researchers, Pierre van Hiele and Dina van Hiele-Geldof (1986), developed a theory for how children grow in geometric thinking. The van Hiele Levels of Geometric Thought provide a five-level progression, the first three of which can occur in elementary school. Just a note: The first level of the model is numbered 0, not because this level is of zero importance. It's because the van Hieles are Dutch and in the Netherlands, they follow the European style of numbering things. For example, when numbering a building's floors, the ground floor often has no number or is assigned the number 0. Therefore, the next floor up is assigned the number 1 and is the first floor, whereas in the United States and Canada, the bottom floor is counted as number 1 or the first floor. The next floor up then becomes the second floor and so on. Aren't numbering systems cool?

2. What geometry ideas are appropriate for preschoolers?

Geometry begins with preschool foundations for shapes and spatial thinking. In addition to naming common shapes including the rhombus, square, triangle, circle, and rectangle, young children need hands-on experience making shapes with dough and drawings. When children make shapes, they begin

"PAUSE"ATIVE BOX

The van Hiele Levels of Geometric Thought

Take a moment to pause and review the following five-level progression of geometric thought (van Hiele 1986). What do you see your child doing that indicates geometric thought?

Level	Explanation
0: Visualization	At this stage, children recognize, name, and classify shapes based on their appearance. For example, a child may recognize a triangle when it is presented with a base oriented horizontally. When the triangle is rotated, however, the child may no longer identify it as a triangle, "because it doesn't look like a triangle."
1: Analysis	Children at this stage of geometric development begin to work with groups of shapes rather than individual shapes. They can start to discuss properties of shapes (like the number or length of sides, the angles, and congruence) rather than just the way a shape looks. At this level, children should still use manipulatives, drawings, and other materials to make sense of the properties of various shapes.
2: Informal Deduction	At this stage, children concentrate on the properties of shapes and relationships between those properties. Up to this point, children may not have been able to see that shapes can be subclasses of one another. For example, all squares are rectangles and all rectangles are parallelograms. Students begin to use if-then reasoning to classify shapes and create and follow informal deductive arguments.
3: Deduction	This advanced level of abstraction takes place in high school geometry courses where students learn to develop and compare proofs and theorems.
4: Rigor	The highest of the van Hiele levels, Rigor, is generally reserved for the college mathematics major who is studying geometry in various systems.

to understand the attributes that make each shape unique. For instance, children can form three-dimensional shapes with dough. When molding a sphere by rolling dough in their hands, children notice that the sphere has no edges or sides, that it rolls, and that it resembles many common objects like baseballs and oranges. Children also need to understand that turning or sliding a shape doesn't change the shape. To demonstrate this, try a silly experiment. Hold up a stuffed animal. Ask the child what it is—a stuffed animal. Then turn the animal upside down. Now what is it? It's still a stuffed animal. Just turning it over doesn't change the object. The same applies to shapes. Repeat the experiment with a triangle, rhombus, and rectangle to drive this point home. Finally, children focus on spatial reasoning when they use positional words like *over*, *under*, *beside*, and *between* to describe the locations of themselves or other objects. This is best explored in the process of everyday play. Who is behind Max in line? Where is the glue located on the shelf? What toy is between the truck and the doll? Learning to identify objects' positions in relation to other objects builds children's spatial orientation and mathematical vocabulary.

TEACHING TIP

Incorporating Movement with Geometry

Learning about geometry, like all other areas of math, should be active—mentally and physically. Kids learn best when their brains and bodies are engaged. Incorporating movement with geometry subtracts stress, adds fun, and maximizes math-positive mindsets. To incorporate movement into your geometry lessons, try these ideas.

- *Construction Flagging Shapes.* Construction flagging tape is used by construction workers to, for example, rope off a new slab of concrete in the sidewalk. It costs only a few dollars per roll and can be found in home improvement stores and online. It isn't sticky (though the name implies this), does not tangle like yarn does, and because it must be seen from a

(continued)

(continued)

distance, flagging tape comes in fun fluorescent colors. You can use it to learn geometry by tying the ends together to make a loop about 40 feet long. Find a wide-open space and have students grab on to the flagging. Give students challenges to make two-dimensional and three-dimensional shapes. This works well in small groups of six to eight students. Talking together about the attributes of the shapes and working cooperatively to create them supports purposeful math-positive vocabulary development. Most of all, it's fun!

- *Hokey Pokey Shapes*. Print out shapes on cardstock, enough for each student to have a set of four shapes—circle, square, rectangle, and triangle. Students stand in a circle with the cards set on the floor beside them. Sing the "Hokey Pokey" tune replacing the body parts with shape names. For example, "You put your triangle in. You put your triangle out. You put your triangle in and you shake it all about. You do the Hokey Pokey, and you turn yourself around. That's what it's all about!" Repeat until all the shapes have been hokeyed and pokeyed. For an added challenge, include three-dimensional shapes as well.

3. In general, what does elementary school geometry involve and what are some ways I can support children's learning of geometry?

Q&A

Geometry is a topic for all ages, prekindergarten through twelfth grade. The National Council of Teachers of Mathematics encourages teachers to provide age-appropriate geometry activities that build on one another. For example, in preschool children learn the names of two-dimensional shapes when they find them in the environment. *Look at the tiles on the floor. They're squares.* One of the geometry objectives in kindergarten is for children to begin to look carefully at the shapes' attributes. *Hey, look at all these shapes. I recognize that one—it's a square. And that one is a rectangle. They look sort of similar, but the square has four sides the same length.* In first grade, children can start to recognize that some attributes are more important than others when it comes to identifying shapes. *I guess it doesn't matter if the triangles are green or yellow. It's more important to count the sides to tell if it's a triangle.* In third grade, students are applying all they've learned so far about shapes' attributes to classify those shapes. They put shapes into groups according to shared attributes. *I'm going to put all the quadrilaterals in a pile. I know they're all quadrilaterals because they all have four sides. It doesn't matter that they are different colors. It doesn't matter that the sides are different lengths. It doesn't matter that the angles are different. Quadrilaterals just need to have four sides.*

"PAUSE"ATIVE BOX

Summary of Big Ideas for Geometry in Kindergarten through Grade 5

Here are some of the big ideas in geometry children will learn in school, including activities you can do to support their learning. The ideas are connected to the Common Core State Standards (National Governors Association Center for Best Practices and Council of Chief State School Officers 2010). Pause and take a moment to review the list. Put an asterisk next to some of the activities you'd like to try. Try them with your child. How did it go?

Grade	Big Ideas in Geometry	Math Activities
K	Name and describe shapes found in the environment. Describe the relative positions of objects using terms such as *above, below, beside, in front of, behind*, and *next to.*	Start a conversation about positional words while putting toys away. "What toy is above the truck on the shelf? What is beside the truck? How would you describe where the box of blocks is?" Using positional words in everyday conversations builds meaning and personal connections to the ideas.

(continued)

(continued)

Grade	Big Ideas in Geometry	Math Activities
K	Correctly name shapes regardless of their orientations or size.	Show shapes in many orientations. For example, there is no "upside down" triangle, just a triangle oriented in a different direction.
	Identify shapes as 2D ("flat") or 3D ("solid").	Use correct terms when naming shapes. For example, while playing with modeling clay, ask, "What shape did you make when you rolled it in your hands? We can call it a ball. Or in math we call it a sphere." Same goes for cans (cylinders) and boxes (cubes or rectangular prisms).
	Analyze, compare, create, and compose shapes.	Make shapes with clay, paper, strings—whatever interests children. Talk about the attributes of the shapes as you play. "What is it that makes a rhombus unique? It has four sides all the same length."
	Analyze and compare two- and three-dimensional shapes using informal language to describe similarities, differences, parts (e.g., number of sides) and other attributes (e.g., having sides of equal length).	How are rectangles, rhombi, and squares alike? They are all quadrilaterals (four-sided figures). How are they different? See if children can come up with at least three ways they are unique.
	Draw shapes and use materials to build shapes (e.g., marshmallows and pretzel sticks).	Use drinking straws and balls of clay to make triangles. The balls are the vertices and the straws are the sides. Don't forget to offer scissors so the triangles aren't all equilateral (same length sides).
	Compose simple shapes to form larger shapes.	Use masking tape to make a shape on a tabletop. It can be a trapezoid (four-sided shape with one set of parallel lines), rectangle, square, or another shape of your choosing. Use pattern blocks or wooden unit blocks to fill in the shape. This shows how small, simple shapes can be combined to create (compose) new shapes.

(continued)

(continued)

Grade	Big Ideas in Geometry	Math Activities
1	Distinguish between defining attributes (e.g., quadrilaterals are closed and four-sided) and non-defining attributes (e.g., color). Build and draw shapes according to their defining attributes.	Play *Secret Shape*. Give children a set of attribute blocks (for more on attribute blocks, see *Manipulatives for Use with Geometry* in this chapter). To start, give clues that include only non-defining attributes, such as *The shape is blue*. Children eliminate all the shapes that "don't fit" the criteria. Next, give defining attributes that apply to more than one shape, for example *It has four sides.* Then provide its defining attributes—those that apply only to one shape, such as *It has four sides. Two are short and two are long.*
	Compose 2D shapes (rectangles, squares, trapezoids, triangles, half-circles, and quarter-circles) or 3D shapes (cubes, right rectangular prisms, right circular cones, and right circular cylinders) to create a composite shape. Compose new shapes from the composite shape.	Sit back to back. One player uses four pattern blocks to create a design, then describes it to the other player, who tries to recreate it using their own blocks. It's harder than it seems and provides great practice with geometry terms.
	Partition circles and rectangles into two and four equal parts (halves and fourths/quarters). Understand that decomposing into more equal parts creates smaller parts.	Use strips of paper to fold halves and fourths. Talk about how the pieces get smaller with more folds.
2	Identify triangles, quadrilaterals, pentagons, hexagons, and cubes.	Draw attention to the many types of triangles by drawing acute, obtuse, equilateral, isosceles, and scalene triangles along with other shapes on a sheet of paper. Have children cross off a triangle and tell how they know it's a triangle. It has three sides and three vertices!
	Recognize and draw shapes with a given number of angles or a given number of equal faces.	Use the corner of a sheet of paper to check if an angle measures 90 degrees. Young children might call these square *corners* while older children use the term *right angles*.

(continued)

(continued)

Grade	Big Ideas in Geometry	Math Activities
2	Partition a rectangle into rows and columns and count to find the total number of squares.	Partitioning a shape into smaller sections or into new shapes helps children "see" how shapes can be composed of other shapes. This idea also supports fraction understanding. Origami is a fun way to play with this idea.
	Partition circles and rectangles into two, three, or four equal parts (halves, thirds, fourths) and recognize a whole as being two halves, three thirds, and so on. Recognize that equal shares of identical wholes can be different shapes.	
3	Understand that shapes in different categories (e.g., rhombi and rectangles) may share attributes (e.g., having four sides), and that the shared attributes can define a larger category (e.g., quadrilaterals). Recognize rhombi, rectangles, and squares as examples of quadrilaterals, and draw examples of quadrilaterals that do not belong to any of these subcategories.	Using reasoning to classify shapes according to a hierarchy takes practice. You might start with something kids connect to—food. Raid the kitchen for two types of fruits (say bananas and apples), a slice of bread, and a fork. What is the most general name you could give these items? They're all things found in the kitchen. Can you give the item a more specific name? The bread, bananas, and apples can be grouped together as foods. The bread is a food, but it isn't a fruit, so it doesn't strictly belong with the apples and bananas. The fork is not a food at all, so it's left out since it doesn't fit the criteria. Using this discussion as a springboard to classifying shapes helps children see how objects can be grouped together in different ways to come up with different categories. Similarly, shapes can be sorted by criteria. For example, a square is a special kind of rectangle as well as a type of rhombus. It's also a quadrilateral. It just depends on what you're sorting by—number of sides (placing it with the quadrilaterals), length of sides (which makes it a member of the rhombus family), or number of sets of parallel sides (which makes it a rectangle).

(continued)

(continued)

Grade	Big Ideas in Geometry	Math Activities
3	Partition shapes into parts with equal areas. Express the area of each part as a unit fraction of the whole.	Geoboards allow for easy partitioning of shapes into equal parts. Make a shape, then use additional geobands (elastic bands) to show the partitions. For example, an equilateral triangle can be partitioned into smaller triangles. How many triangles will fit? Try it and find out.
4	Identify and draw points, line segments, rays, angles (right, acute, obtuse), and perpendicular and parallel lines.	Geometry terms can be, well, obtuse. Keeping them straight is tricky because they aren't used very often outside of the classroom. It might be helpful to make a visual reminder. A pyramid diorama is a graphic organizer with a section for each term. To make a diorama, fold a square sheet of paper in half to make two triangles. Cut along the fold line to the exact center of the square. Then fold the pieces over and glue or tape them to make a triangle that stands up. Repeat four times and tape together to make a pyramid-like shape. For each of the geometry terms, have children draw a picture and give a simple definition and/or a real-world example. For instance, for the term *parallel lines*, draw a train track and write a child-friendly definition like, "two lines that never meet."
	Classify 2D figures based on their lines and angles. Recognize right triangles as a category.	David A. Adler's book *Triangles* (2015) gives entertaining explanations about all the different types of triangles. Read it while eating a triangle-shaped snack (like watermelon slices) for a bit of yummy math-positive learning.
	Identify and draw lines of symmetry for 2D figures.	Using an unbreakable mirror (sold commercially as a Mira) helps students see angles and symmetry in designs made with pattern blocks.

(continued)

(continued)

Grade	Big Ideas in Geometry	Math Activities
5	Graph ordered pairs of numbers as points on a coordinate system (e.g., *x*-axis and *x*-coordinate, *y*-axis and *y*-coordinate).	The game *Hurkle* (Stenmark, Thompson, and Cossey 1986) provides an introduction to ordered pairs and the coordinate plane system. Try it with your family by searching for it online and printing out the accompanying game board.
	Represent and interpret real world and mathematical problems by graphing points in the first quadrant (positive *x* and positive *y*) of the coordinate plane.	
	Classify 2D figures according to their properties.	Venn diagrams (intersecting circles like below) provide a visual space for sorting shapes. Use rectangles, squares, rhombi, and other quadrilaterals to explore the classification of shapes according to attributes.
	Understand that attributes belonging to a category of 2D figures also belong to all subcategories of that category (e.g., rectangles have four right angles and squares are rectangles, so all squares have four right angles).	
	Classify 2D figures in a hierarchy based on properties.	

4. What do some common geometry terms mean?

Vocabulary is key to communication about geometric ideas. The language of mathematics, and geometry in particular, can be tricky. I recommend purchasing a math dictionary written for children and keeping it close by as a handy reference. The definitions I give in this resource come from *Math Dictionary: The Easy, Simple, Fun Guide to Help Math Phobics Become Math Lovers* (Monroe 2006). You can also find online math dictionaries with simplified definitions of common mathematical terms. Here are some terms you might come across in children's geometry homework.

"PAUSE"ATIVE BOX

Common Geometry Terms

Take a moment to pause and review the following terms. Then do a *Geometry Walk* (see the activity that follows this table) with your child.

Term	Definition	Example
Two-Dimensional Shapes	Two-dimensional shapes lie in a flat plane. Sometimes they are called *plane figures*. Two-dimensional shapes do not have thickness. You could not open them up and pour chocolate chips into them (or healthy food, for that matter). They have two dimensions we can measure, length and width.	Circles, rectangles, squares (a special type of rectangle), rhombi, parallelograms, trapezoids, hexagons, pentagons, octagons, ovals, triangles, half-circles, and quarter-circles are all 2D shapes.
Polygons	These are closed two-dimensional figures formed by joined line segments.	Triangles, quadrilaterals, pentagons, and hexagons are all examples of polygons.
Congruent	*Congruent* figures have the same size and shape.	A scale model maintains congruence. It's enlarged or shrunken down but doesn't change the shape's proportionality. That means the shapes are congruent.
Sides	*Sides* of a polygon are the line segments that form the shape.	A quadrilateral has four sides. A triangle has three sides. A circle has no sides.
Width	*Width* is usually thought of as the distance from side to side. In math, there is some agreement that width is the shorter of the two sides on a rectangle.	The width of a classic Dr. Seuss hardbound book is 6.9 inches.
Height	*Height* is usually thought of as how tall a person or thing is.	The height of an average 8-year-old child is 50 inches.
Vertex	A point at which two or more sides or edges of a figure meet. Also called a *corner*. The plural term for vertex is *vertices*. I know, it's confusing. But speaking with precision helps everybody know we're all talking about the same part of a shape.	A triangle has three sides and three vertices.

Term	Definition	Example
Three-Dimensional Shapes	Three-dimensional shapes, sometimes called *space figures*, have thickness. You could open them up and pour chocolate chips inside. They have three dimensions we can measure—length, width, and height.	The 3D shapes typically introduced in elementary school include: cone, cube, cylinder, rectangular prism, triangular prism, sphere, and pyramids (named for the shape of their base).
Polyhedron	A geometric solid whose faces are all identical, regular polygons meeting at the same three-dimensional angles.	The tetrahedron (or triangular pyramid), cube, octahedron, dodecahedron, and icosahedron are regular polyhedrons.
Faces	*Faces* are the closed figures that form the individual surfaces of a space figure. Not all space figures have faces. For instance, a cube has six square faces, but a sphere has no faces.	While the faces of polyhedrons are all identical, the faces of a rectangular prism are not. Two of them are squares and four of them are rectangles.
Edges	*Edges* are formed where two faces of a 3D shape meet.	An icosahedron has 30 edges. Now that's a polyhedron to commit to memory.
Triangles	*Triangles* are named by their sides or their angles.	*Equilateral triangles* have three sides of equal length. *Isosceles triangles* have two sides of equal length. *Scalene triangles* have no sides of equal length. *Right triangles* have one 90 degree angle. *Obtuse triangles* have one angle greater than 90 degrees, and two angles that measure less than 90 degrees. *Acute triangles* have angles measuring less than 90 degrees.
Tiling	*Tiling* is a pattern of shapes repeated to fill a plane with no gaps or overlapping. Sometimes we call this a *tessellation*.	Many modern art pieces, quilts, and coloring books contain tessellations. Search online to find pieces by M. C. Escher, a master at creating mind-boggling tessellations that show geometry is the building block of his creativity.

"PAUSE"ATIVE BOX

Geometry Walk

It can be fun to take a *Geometry Walk* around your home looking for circles, triangles, and rectangles, including squares, as you describe the shapes of sofas, bookshelves, laundry hampers, and dishes. This activity helps children recognize shapes and their functionality in the real world. All you need is a sheet of paper, a pencil, and a keen eye.

Instructions

1. Look around the house. What shapes do you see? For example, look at the frame around the front door. What shape is it? Why do you think the door opening is a rectangle? Would a circle be a good shape for a door? Why or why not?
2. Look at the toilet paper roll. What shape is it? Why is a cylinder a handy shape for a roll of toilet paper? Would another shape work as well?
3. On a sheet of paper, draw this table. Go on a geometry walk to fill in the table.

Object	Shape	Why the Shape Makes Sense for This Object

4. What is the most interesting shape you found and why do you think the shape works well for that object?

Teaching Geometry Terms

TEACHING TIP

Geometry terms can be confusing, partly because children don't have much exposure to them outside of math class. Teachers must, therefore, find strategies to build children's vocabulary in meaningful ways.

Teach:

- intentionally (make time to introduce, explore, and review terms)
- purposefully (use your curriculum materials or state objectives as a guide)
- engagingly (use games, lessons, and activities to make learning fun)
- by integrating vocabulary into the body of a math lesson (when it makes sense)
- by reviewing terms often (create a math word wall—better yet, have children create and maintain it); read more about word walls in the Teaching Tip in Chapter 6, Question No. 4, *What do some common measurement terms mean?*
- by modeling the correct use of terms (always aim for accuracy in your own language)
- by scaffolding children's developing mathematical vocabularies (reinforce attempts during discussions and bridge from informal to formal language)

5. What is the difference between a kite, a diamond, and a rhombus?

We sometimes adopt an informal term for a geometric figure:

- *Diamond:* A diamond is a good example of this confusion. A diamond is *not* a geometric figure. You can almost never go wrong giving it as a gift, but it is a gemstone, not a shape. The term *diamond* has been adopted as an informal term for the rhombus shape, so let's see if we can set the record straight.
- *Rhombus:* A rhombus is defined as a four-sided figure with all sides the same length (congruent).
- *Kite:* A kite is a geometric shape but is not usually introduced in elementary school. A kite is a four-sided figure with two sets of congruent sides.

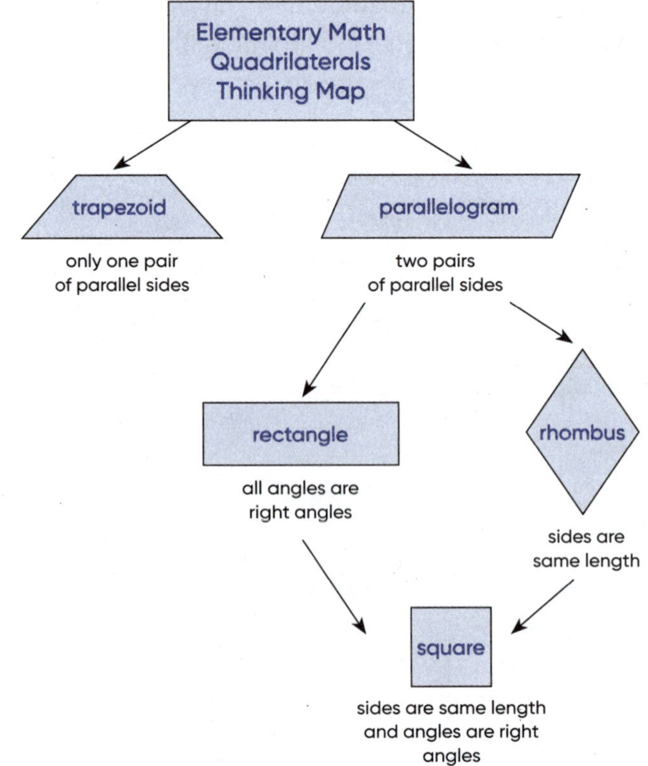

Using these definitions, we see that a rhombus is a handy shape. A square is a special rhombus in which all angles are right angles. A parallelogram, if it has four sides the same length, is a special rhombus. Very young children do not necessarily need to know about the hierarchical nature of these relationships between shapes. But we should use the correct terminology from the start so that when the time comes, children can connect new understanding to prior learning.

Connecting Geometry to the Real World

TEACHING TIP

Geometry can seem like a subject reserved only for the classroom unless we open children's eyes to the shapes, angles, sides, and faces staring back at us. Teachers can give students geometry eyes by sending them on a *Geometry Scavenger Hunt* with digital cameras or tablets to snap photos of geometric discoveries in their school, playground, and neighborhoods.

As an engaging introduction, you might read the classic book *Shapes, Shapes, Shapes* (Hoban 1996). The book uses photographs to document the geometry in the buildings, bridges, and playgrounds in an urban area. Give students a list of geometry terms and shapes that are age-appropriate. For example, kindergarteners may search for two-dimensional shapes while fourth graders can photograph two- and three-dimensional shapes and all sorts of angles. Students can share their photos with the class by printing them out for a class book or creating a digital presentation. When children recognize the functionality of shapes in our environment, they develop a greater appreciation for the wonder and beauty of geometry and build math-positive mindsets that set them up for future success.

6. What do composing and decomposing have to do with shapes?

Composition is a big idea across many strands in mathematics. It means that whole things can be made up of smaller things. Chapter 4, Question No. 6 (*A homework assignment asks my child to solve using number bonds. What is a number bond?*), describes how numbers can be composed of smaller numbers—like the number 5 being composed of 4 + 1 and 2 + 3, for example. The same goes for shapes. When we fold a square in half diagonally, we see how the square could be composed of two triangles. When we fold it horizontally, we see how the square could be composed of two rectangles. Exploring this concept starts early. Young children play with composition as they build towers and designs by combining two- and three-dimensional shapes in the block area of the classroom. They also construct shapes with balls of clay and wooden craft sticks or plastic straws. Older children begin to understand how two-dimensional shapes combine to form three-dimensional shapes. They do this by stamping a three-dimensional shape into a flattened piece of dough, discovering that the faces of a cube, for example, are all squares.

TEACHING TIP

Composing Shapes in the Block Center

Blocks are a mainstay of early childhood classrooms. Ideal for developing concepts of cause and effect, stability, and engineering, blocks also shape-up nicely for learning geometry. How can you leverage block play to include more elements of composition? One suggestion is to take a photograph of students and their block structure at the end of their play. This serves two purposes. First, it celebrates students' creativity and spatial reasoning before the structure must be "decomposed" and put away. Second, if the teacher prints out the photographs and compiles them in a class book, students can refer to the book and try to re-construct (compose) their former design. This is harder than it appears! Students must count blocks, compare them to select the correct size and shape, even make assumptions about blocks that aren't visible in the photo. Now that is geometric reasoning at work!

7. What are some materials and activities to help children understand geometry?

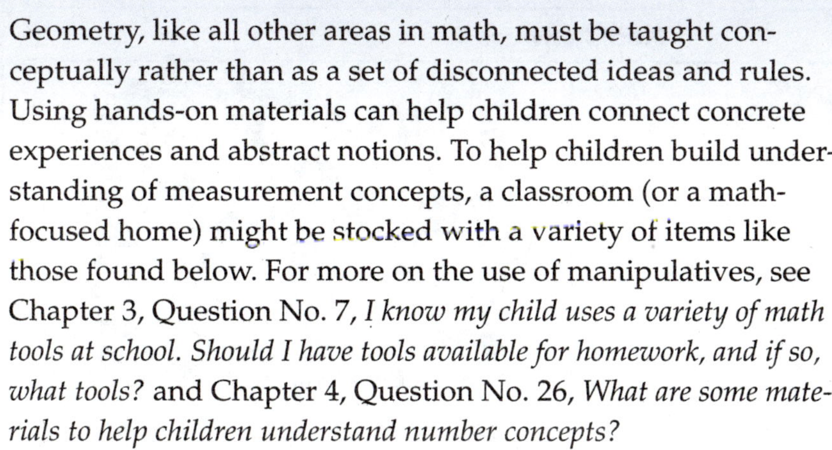

Geometry, like all other areas in math, must be taught conceptually rather than as a set of disconnected ideas and rules. Using hands-on materials can help children connect concrete experiences and abstract notions. To help children build understanding of measurement concepts, a classroom (or a math-focused home) might be stocked with a variety of items like those found below. For more on the use of manipulatives, see Chapter 3, Question No. 7, *I know my child uses a variety of math tools at school. Should I have tools available for homework, and if so, what tools?* and Chapter 4, Question No. 26, *What are some materials to help children understand number concepts?*

"PAUSE"ATIVE BOX

Manipulatives for Use with Geometry

Below is a list of manipulatives most frequently stocked in classrooms for supporting the learning of geometry. Take a moment to pause and review the list. I've also given suggestions for simple geometry tasks using each type of manipulative. Put an asterisk by the activities you'd like to try with your child. Try them out. How did it go? Have fun building a math-positive mindset about geometry along with your child.

Manipulative	**Description**	**Uses in Geometry**
Pattern Blocks	Plastic or wooden blocks that come in six shapes: yellow hexagons, orange squares, red trapezoids, green triangles, tan rhombi, and blue rhombi. 	• Attributes such as color, shape, and number of sides • Sorting and classifying • Composing and decomposing shapes (for example, six green triangles make a yellow hexagon) • Symmetry • Spatial reasoning

Math Activities

Play *Look, Make, Fix* (Copley 2010). First build a design using three or four pattern block pieces. The pieces must be touching. Have children carefully study your design. This is the *Look* phase. Then cover the design with a sheet of paper and have children try to *Make* it. Then lift the paper off and have children *Fix* their design, if needed. Trade jobs. This activity builds spatial reasoning as well as ideas about how shapes can be combined to make new shapes (shape composition).

In a game of *Look, Make, Fix*, my son Duncan attempted to replicate my design. What do you notice about the two designs? How are they similar and different?

(continued)

(continued)

Manipulative	Description	Uses in Geometry
Attribute Blocks	Most sets of attribute blocks include several shapes (triangle, square, rectangle, circle, and hexagon) in two sizes, two thicknesses, and four colors.	• Analyzing and comparing shapes to describe how they are similar and different • Sorting and classifying

Math Activities

Play *Attribute Block Trains*. Build a train. Decide first if you want your train to be a 1-, 2-, or 3-difference train. For example, if you choose to make a 2-differences train this means each block will be different from the one immediately preceding it in exactly two ways. Since thickness and color are difficult to show in black-and-white photographs, the other two attributes are used in the examples. Look at the image to the right showing the 1-difference train. The first shape is a large yellow circle. The next shape (a small yellow circle) is different in one way—size. How is the third shape (a small red triangle) different from the second shape? Each successive "train car" must be different in one way from the car before it. Now look at the 2-differences train.

The large yellow square is followed by a small, yellow triangle which is different from the first shape in two ways—size and shape. Next is a large yellow rectangle. It is different from the small yellow triangle in two ways. It is a different shape and a different size. Don't forget you can make 3-differences trains, too. *Attribute Block Trains* requires players to talk about the attributes of the shapes as they make sure each piece fits the criteria. Play it several times. Is it easier to make a 2-differences train than it is to make a 3-differences train? The answer might surprise you.

(continued)

(continued)

Manipulative	**Description**	**Uses in Geometry**
Geoboards	A plastic board with pegs, around which children wrap rubber bands to form shapes.	• Composing and decomposing shapes (for example, wrapping a band two across and three down to make a rectangle) • Area (count the square units inside the shape to find the area) • Perimeter (count the units along the outside edges of the shape to find the perimeter) • Identifying attributes such as sides, vertices, and angles

Math Activities

Play *Secret Shape*. Sit back-to-back with your child. Each of you needs your own geoboard and elastic geobands. Stretch the geobands between pegs to make a shape. Describe the shape to your child and have them try to copy it. For a younger child, you might say, "My shape has five sides all the same length."

This could look like the first shape shown here—a regular pentagon. If you have an older child, you can use regular and irregular shapes. "My shape has four sides and only one right angle." This might look like the second or third shapes shown in the photo—though there are many shapes that fit the criteria. This activity builds vocabulary and concepts about the defining attributes of shapes including sides, vertices, and angles.

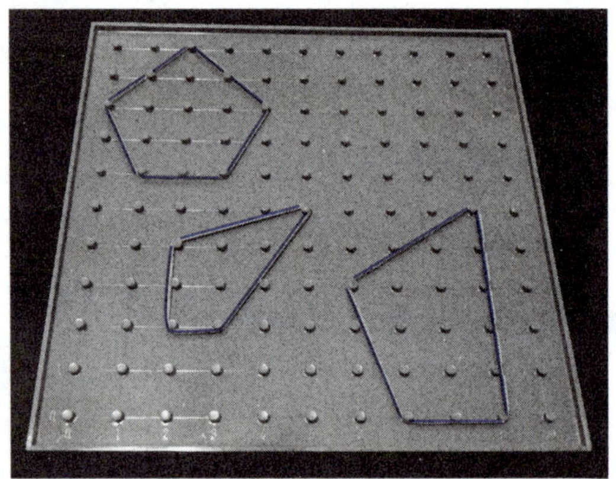

This geoboard has three secret shapes in play. Two are quadrilaterals and one is a pentagon.

(continued)

(continued)

Manipulative	Description	Uses in Geometry
Geometric Solids	Plastic or wooden sets often include cones, prisms, cubes, spheres, cylinders, and pyramids.	• Attributes of shapes such as edges, faces, vertices, and angles • Shape composition and decomposition • Sorting and classification

Math Activities

Play *Block Tower Building*. Have a contest to see who can build the tallest tower. Children toss a die and build a tower with the indicated number of blocks. Then it's the grown-up's turn to start a tower. After each player has had three turns, compare the heights of the towers. This activity gets children thinking about the attributes of 3D shapes.

Talk about what types of blocks work well for building towers and which do not. "Which of the blocks stack well? Which blocks roll? Which have points and which have flat sides?" When you play next time, add an engineering design challenge by seeing who can build a tower that withstands the force of a small table-top fan. Understanding geometry opens a world of math-positive possibilities from engineering and construction to architecture and art.

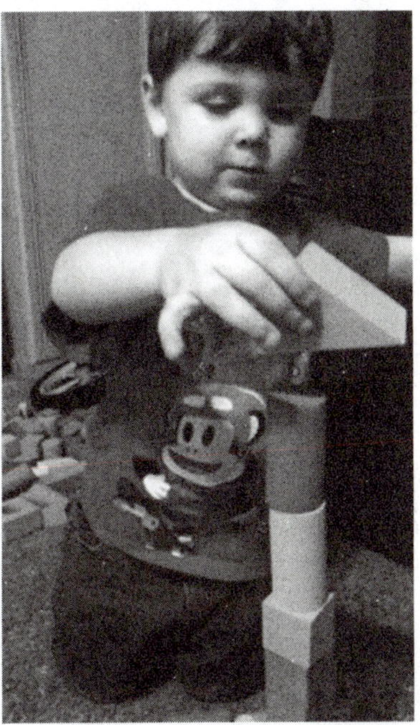

When he was three years old, my son Quinn loved *Block Tower Building*. He learned to say, "Oh, well!" and shrug when a tower toppled over. Then he began building once again. That's a math-positive mindset!

(continued)

(continued)

Manipulative	Description	Uses in Geometry
Tangrams	A tangram is a traditional Chinese geometric puzzle consisting of a square cut into seven pieces, called tans, that can be arranged to make other shapes and designs.	• Shape composition and decomposition • Spatial reasoning • Relationships between shapes

Math Activities

Play *Tangram Animal Tales*. Help children use the tangrams to make animals like those shown in the figure here. Have children make up a story about the animals. You might record your child telling the story and moving the tangram pieces around to make different animals.	 In a game of *Tangram Animal Tales*, my son Chas made a turtle and a fox for his story.

(continued)

Manipulative	Description	Uses in Geometry
Magnetic Tiles	Translucent, brightly colored plastic shapes that connect using magnets so that children can construct flat or tall structures.	• Shape composition and decomposition • Sorting and classification • Spatial reasoning • Composing and composing shapes

Math Activities

Play *Build a Tile Village*. Magnetic tiles have strong enough magnetic pull to maintain their structure well, so they are ideal for creating a little village of triangular prism houses and rectangular prism stores. Children might want to use a small fridge magnet to hand-make signs identifying the buildings as a grocery store, post office, or house. Talk about how the magnetic shapes fit together to make a new, 3D shape. For example, in the picture here, six small squares form a small cube.

The large cube is made of large and small squares used in combination. The rectangular road is made of squares and triangles. This idea of composition is very important in mathematics—not only in shapes but also in combining numbers to make new numbers.

My son Knox built triangular prism houses and a big cube-shaped hotel connected by a sidewalk.

TEACHING TIP

Supporting Reasoning about Spatial Orientation

One of the most overlooked elements of geometric reasoning is spatial orientation. Spatial activities help children begin to see how their bodies relate to the things in their environment. Try these activities to support spatial orientation with your children.

- *Spaceman Says:* This is like Simon Says only no one gets eliminated from play and the Spaceman refers to orientation in a space. Give each child a stuffed animal and provide directions using positional words like *over*, *under*, *beside*, *below*, *between*, and *behind*. A child can be the leader or the teacher can give the directions. For example, "Spaceman Says put your animal under your arm." "Spaceman Says put your animal below your chair." "Spaceman says put your animal on your head." Active, silly, spatial thinking!
- *Positional Word Obstacle Course:* Use crepe paper to create this obstacle course. Children use positional words to describe how they move through the course, going over, under, between, beside, and so on. Friends can watch and tell the positional words too.
- *Rosie's Walk:* Use the classic picture book *Rosie's Walk* (Hutchins, 1967) as a context for practicing positional words. Work as a class to create a map of Rosie's walk (she's a chicken) through the farmyard. Write the positional words on the map and allow children to use a stuffed animal chicken to retell the tale.

QUICK REFERENCE CHART

Common Concerns about Geometry

This table is a quick reference for some of the *most common concerns* I hear from parents. The table lists the concern, an explanation of the thinking behind it, and an idea that addresses the concern. Consider photocopying this table and placing it somewhere where you can be reminded of what is needed to build your math-positive mindset and that of your children and/or students.

Concern	Explanation	What to Do	Question Number(s) (to Learn More)
Why teach geometry in elementary school?	Geometry includes the properties of shapes, our orientation to objects in space (*between, above, below,* etc.), and symmetry, transformation, and proportionality. Advancing through the levels of geometric thought moves children along a continuum of increasingly abstract geometric thinking.	Take a geometry journey in your kitchen by naming shapes, identifying their attributes, then classifying shapes by their common attributes. This is especially fun if you use real objects like bowls, boxes, and cans.	1
Learning shape attributes	Child must recognize that rotating a shape doesn't change its name or its attributes.	Go on a *Shape Hunt* with children to identify 2D and 3D shapes. Use sticky notes to mark the squares, index cards for rectangles, dot stickers for circles. What could you use to mark the triangles? Ask children for ideas.	2
Supporting geometry understand-ing at home	Geometry shouldn't be reserved for high school proofs and theorems. At home, it can be about the contents of the kitchen cupboard and rhombi (diamonds) twinkling in the sky.	Shape up your math-positive mindset for geometry by trying out some of the games, activities, and conversation starters listed in this chapter. Before you know it, children will be angling to do more geometry.	2

(continued)

(continued)

Concern	Explanation	What to Do	Question Number(s) (to Learn More)
What are geometry concepts in elementary school?	Elementary school geometry focuses on shapes, their attributes, and ways to classify them into groups.	Try this triangle game: Draw two dots on a card. Your child draws a third dot then connects the dots to form a triangle. Try it again drawing the first two dots in the exact same spots as last time. Will your child make a different triangle by placing the third dot in a new spot? I bet so! Triangles can look very different from one another but still meet the definition of a triangle because they have three sides.	3
Where can I find child-friendly definitions of geometry terms?	Geometry terms can be tricky, particularly since many of them are not used much in everyday conversation or in other subjects in school.	A handy child-friendly math dictionary can provide simple definitions of geometry terms and illustrations. Check out *Math Dictionary: The Easy, Simple, Fun Guide to Help Math Phobics Become Math Lovers* (Monroe 2006).	4
What's a rhombus?	Four-sided figure with all sides of the same length. Incorrectly called a diamond.	Challenge children to find errors in picture books and on kids' TV shows.	5
How do composing and decomposing relate to geometry?	Just like a number can be composed of other numbers (like 5 being composed of 4 and 1), shapes can be composed of other shapes. A square can be composed of two triangles, for example.	Tangrams are great for seeing how shapes can be composed of other shapes. Try making a square using two shapes or four shapes. What do you notice? Exploring with shapes helps build awareness of shape composition.	6
What geometry tools should I have on hand?	Geometry is a hands-on endeavor. Using manipulatives helps children build ideas about shapes, attributes, and geometry terms.	Pattern blocks, attribute blocks, geoboards, geometric solids, tangrams, and magnetic tiles all make learning geometry fun and meaningful.	7

I Can Help with Measurement and Data

CHAPTER 6

(continued)

(continued)

TEACHING TIPS

1. Why measurement? What does measurement in preschool and elementary school look like?

Measurement lends context to many of the other areas in mathematics. We measure our bodies, our time, and our bank accounts using number. We measure shapes' side lengths and angles in geometry. We use measurement to compare bar graphs to determine which survey response was most popular. In these ways, we use measurement to guide our math-positive explorations of many mathematical ideas. Measurement also connects to other school subjects. For example, we measure our heart rates in physical education and measure outcomes of science experiments. Simply put, measurement is assigning a numerical value to a measurable attribute. It makes sense then that young children first must understand that objects have attributes that can and cannot be measured. Some non-measurable attributes include scent, color, and texture. Some measurable attributes include:

- Time
- Temperature
- Money (value)
- Length
- Area
- Volume (capacity)
- Weight (mass)

Preschoolers' initial ideas about measurement should focus on direct comparisons between two objects. For example, they may stand back-to-back to see who is taller. Comparisons should be number-free initially (e.g., longer, shorter, heavier, lighter, faster, slower) and involve only two objects for comparison.

Next, children need to begin to see the need for a "unit" of measurement. They use their counting and comparing skills along with an appropriate nonstandard unit of measurement. Paper clips and drinking straws are good for measuring length. Flat-sided glass marbles work well for measuring weight on a pan balance. Measure capacity using scoops of rice or dry beans. Children won't spend their lives measuring how many shoes-long a basketball court is, but it helps them initially understand how measurement and counting work together.

Eventually, the need for standard units should be developed. Inches, pounds, and miles are all examples of customary units. Centimeters, grams, and kilometers are examples of metric units of measurement. Children should also be able to identify and know when and how to use thermometers, clocks, and coins.

Finally, with abundant experience measuring using standard units, children begin to see the need for formulas. Formulas take all the previous experiences with measurement and shorten things up. Children can apply their ideas of measurement to see patterns, generalize, and derive formulas for measurement applications such as area, perimeter, and volume.

2. What are the big ideas in elementary school measurement?

Measurement is an active experience with hands-on tasks building ideas about length, weight, capacity (volume), time, and money. Children are front and center as doers of mathematics, engaged in choosing objects to measure, tools to measure with, and units (like inches or centimeters) to report. See the table of big ideas for measurement starting on page 224.

TEACHING TIP

Teaching Measurement Well

I've often thought teaching all content in mathematics through a measurement lens would allow for authentic, real-world applications of almost all elementary school mathematics concepts. Measurement truly is among the strongest real-world connections in mathematics. To teach measurement (and therefore mathematics) well, I recommend keeping the following in mind:

- *Real-World Scenarios:* Be certain all activities are positioned in "real-world" scenarios that speak to children's interests and experiences.
- *Benchmarks:* When we help children develop benchmarks for standard units, like the width of their finger being about one centimeter or the length from fingertip to elbow being about one foot, they have handy references for abstract mathematical ideas. For more on benchmarks in measurement, see Question No. 7 in this chapter.
- *Open-Ended Problems:* Measurement must always be taught using hands-on, math-positive problem situations that allow for productive struggle and multiple avenues for finding an answer. Students must be actively doing, experimenting, and performing—not passively observing or filling in worksheets.
- *Estimation:* Wherever possible, encourage children to estimate prior to measuring and revisit their estimate when they are halfway through to revise it if necessary. This builds number sense and helps children develop reasoning skills along with a math-positive mindset. For more on the importance of estimation in measurement, see Question No. 6 in this chapter.

"PAUSE"ATIVE BOX

Summary of Big Ideas for Measurement for Kindergarten through Grade 5

Here are some of the big ideas for measurement that children will learn in school, including activities you can do to support their learning. The big ideas are connected to the Common Core State Standards (National Governors Association Center for Best Practices and Council of Chief State School Officers 2010). Pause and take a moment to review the list. Put an asterisk next to some of the activities you'd like to try. Try them with your child. How did it go?

Grade	Big Ideas In Measurement	Math Activities
K	Describe measurable attributes such as length or weight of an object.	Start a conversation: Make a list of all the attributes of a button—its color, size, texture, weight, number of holes, and so on. Which of these attributes can be measured?
	Directly compare two objects to see which object has more of/less of a common attribute.	Use a pan balance to compare the weights of two small objects like a glue bottle and a glue stick.
1	Order three objects by length.	Have children choose three toys. Use strings to measure the length of the toys and order the strings from longest to shortest. Which one is closest to the length of a child's shoe?
	Compare the lengths of two objects by using a third object.	
	Lay multiple copies of a shorter object end to end to find the length of a longer object.	Have children choose an object to measure and a nonstandard unit like paper clips. Find the length of the object by laying the paper clips end to end and counting the paper clips.
	Tell and write time in hours and half-hours using analog and digital clocks.	Play the *Catch the Clock* game. To do so, make up a silly dance to celebrate when children notice the clock showing the hour or half hour.

(continued)

(continued)

Grade	Big Ideas in Measurement	Math Activities
2	Select and use appropriate tools such as rulers, yardsticks, meter sticks, and measuring tapes.	In previous grades, children used many nonstandard units, some of which approximated standard units like inch cubes or centimeter cubes. Connect these familiar tools to the units shown on rulers and measuring tapes. For example, measure first with inch cubes then with a ruler and compare the results.
	Measure the length of an object twice, using units of different lengths for the two measurements.	First use white Cuisenaire rods then use the orange rods (ten times as long as the white) to measure the length of an object. Ask children what they notice. Discuss.
	Describe how measurement relates to the size of the unit used.	
	Estimate length using inches, feet, centimeters, and meters.	Trace your child's hand on a sheet of paper. Explore the marks on a ruler and find which indicate centimeters. Have your child estimate the width across the traced hand in centimeters, then measure to check. Repeat with the other hand. Ask, "What do you notice?" Now, trace an adult's hand. Ask your child, "How can your previous measuring help you estimate the adult's hand? Do you think it will be larger, smaller, or the same? Why?"
	Measure to determine how much longer one object is than another.	Take a nature walk to gather items, then have children use a ruler to find each item's length. Ask children to order the items from longest to shortest and tell how they know.

(continued)

(continued)

Grade	Big Ideas in Measurement	Math Activities
2	Use addition and subtraction within 100 to solve word problems involving lengths.	Try this word problem with children: *Knox is 43 inches tall. When he was born, he was 20 inches long. How many inches has he grown?* This is a fun problem because it shows how length (horizontal) and height (vertical) are two different forms of linear measurement. Since babies can't stand up, we measure their lengths. As soon as children can stand, we measure their heights.
	Represent whole numbers as lengths from 0 on a number line. Represent whole-number sums and differences on a number line.	Take two equal lengths of string and use clothespins to label them String A and String B. Tape String A to the wall and label one end with a 0 and the other end with a 12. Extend String B the full length of String A and talk with children about how the strings are the same length. Then fold String B in half and talk about what number would go there on String A. Mark that spot with a 6. Fold String B in half again and talk about what number would go there on String A. Mark that spot with a 3. Ask, "What other numbers can you figure out?"
	Tell and write time from analog and digital clocks to the nearest five minutes using a.m. and p.m.	Make a timeline of children's daily schedules. Draw pictures and write the correct time for each task.
	Solve word problems involving dollar bills, quarters, dimes, nickels, and pennies using $ and ¢ symbols.	Try this word problem with children: *Anusha wants to buy a snack that costs 50 cents. If the vending machine does not take pennies, what coin combinations could she use?*

(continued)

(continued)

Grade	Big Ideas in Measurement	Math Activities
3	Tell and write time to the nearest minute and use models to solve word problems involving addition and subtraction of time intervals.	Search online for interactive clocks children can use to solve the following scenarios: • *If it is 7:00, what time will it be in 2 hours?* • *If it is 9:00, what time was it 6 hours ago?* • *If it takes 5 hours to get to our cousin's house and we need to be there by noon, what time should we leave?*
	Measure and estimate liquid volumes and masses of objects using grams, kilograms, and liters.	*Three Act Tasks* are short video clips with real-world math modeling sequences. Search online for UHouston Math Coke Cups Three Act Task. This 90-second video gets kids wondering, *How many drink cups can be filled with one 2-liter bottle of soda?* What other Three Act Tasks can you find online for measuring volume? Try a few together!
	Use models and +, −, ×, and ÷ to solve one-step word problems involving masses or volumes.	
	Measure the area of a rectangle by counting unit squares on grid paper or by tiling (covering the rectangle with square units) to show that the area is the same as length times width.	Try this activity together to build math-positive mindsets about area—the number of square units that cover an object: Guess how many one-inch square tiles will cover a book, magazine, and sheet of newspaper. Measure by covering the items with tiles. Keep the tiles on the items and use a ruler to measure the length and width of the items. How do these measurements relate to the area? Use the formula for area (length × width) to figure the area. Where can you "see" this number in the tiles?

(continued)

(continued)

Grade	Big Ideas in Measurement	Math Activities
3	Solve real-world and mathematical problems involving perimeter. Find the perimeter given the side lengths and given an unknown side length. Show how rectangles can have the same perimeter and different areas or the same area and different perimeters.	The distributive property can be used to figure out real-world area problems. For example, my kitchen is too small for our whole family to cook meals together. We want to remodel. If the current kitchen is a 12-foot-by-10-foot room and has an area of 120 feet, how many square feet will we add if we increase the length of one wall by 6 feet? Which wall should we extend to get the most increase? To solve using the distributive property: $a(b + c) = (a \times b) + (a \times c)$ $10(12 + 6) = (10 \times 12) + (10 \times 6)$ Answer: 180 square feet $12(10 + 6) = (12 \times 10) + (12 \times 6)$ Answer: 192 square feet Solve the kitchen remodeling problem then compare the perimeters of the two possible room additions (12 ft. x 16 ft. and 10 ft. x 18 ft.). What do you notice about the area of each room and the perimeter?
4	Know relative sizes of standard units (e.g., how km relates to m and cm) and make conversions between the units.	Play *Silly Units*. Make up silly stories and have children tell which unit of measurement (inches, feet, or yards; or centimeters, meters, or kilometers) would make the story the silliest and why. For example, *I rode my bike 10 (centimeters, meters, or kilometers) to my friend's house. I used 6 (inches, feet, yards) of ribbon to wrap my mother's gift.*

(continued)

(continued)

Grade	Big Ideas in Measurement	Math Activities
4		Make sure children tell why the units are silly. For example, riding a bike 10 centimeters wouldn't even get you out of the garage! Using 6 yards of ribbon to wrap a present would make it pretty tricky to open!
	Use +, −, ×, ÷ to solve word problems involving distance, time, volume, mass, and money, including problems with fractions or decimals.	What word problems can you and your children make up using items in the refrigerator? Here are a few I thought of: *There are four open 32-ounce bottles of ketchup in the fridge. If each of them was full, how many ounces of ketchup would that be?* (And why in the world are there four open bottles of ketchup in the first place?) *String cheese costs $12.19 for a 60-count package. If the Cutler kids eat 8 sticks per day, how much are we spending on string cheese per day? How many days until we run out of string cheese? Why is string cheese so hard to open?* (Okay, now you know my pet peeves about ketchup and cheese. But the math is fun!)
	Use formulas for area and perimeter of rectangles in real-world and mathematical problems.	Here's a spatial thinking problem for area and perimeter: *The sidewalk in front of the school is crowded with kids, bikes, teachers, and parents. If the sidewalk currently measures 30 feet by 8 feet and the principal wants to increase the area of the sidewalk to 320 feet, what are some possible lengths and widths that would work? Which dimensions do you think would work best?* (See Cutler 2007 for many more engaging measurement problems.)

(continued)

(continued)

Grade	Big Ideas in Measurement	Math Activities
4	Recognize angles as geometric shapes that are formed wherever two rays share a common endpoint.	*Turtle Pond*, an interactive activity on the National Council of Teachers of Mathematics website, uses simple computer programming commands to move a turtle through a maze. Try it out with children for a fun introduction to 15-, 30-, 45-, 60-, 75-, and 90-degree angles.
	Solve addition and subtraction problems to find unknown angles on a diagram in real-world and mathematical problems.	
5	Convert between units within a given measurement system (e.g., convert 5 cm to 0.05 m), and use these conversions in solving multistep, real-world problems.	Have children look around their classroom or home and find objects that are the following lengths: 5 cm, 10 cm, 1 m, and 3 m. Write the lengths and names of the objects on separate cards and mix them up and place them face down on the table. When children find a correct match, have them suggest an equivalent representation using a different unit. For example, if you find the desk and 1 meter cards, that's a match and the length could also be described as 100 centimeters or $\frac{1}{1000}$ of a kilometer.
	Recognize volume as an attribute of solid figures and measure volume by counting unit cubes, using cubic centimeters, cubic inches, cubic feet, and nonstandard units.	From the internet, print out four sheets of 1 cm grid paper. Cut the sheets into 12×12 grids. Have children cut one square off each corner and form an open-top box by folding and taping the corners. This box has a capacity of 100 centimeters because 100 centimeter cubes can fit inside this $10 \times 10 \times 1$ centimeter cubed box. Using the same size grid and cutting off different sizes of squares from the corners, ask children, "What is the largest capacity box you can make? How could you figure the capacity without filling the box with cubes?"
	Relate volume to multiplication and addition and solve real-world and mathematical problems involving volume.	
	Find the volume of a right rectangular prism with whole-number side lengths by packing it with unit cubes. Relate this to formulas: $V = l \times w \times h$ and $V = b \times h$	

(continued)

(continued)

Grade	Big Ideas in Measurement	Math Activities
5	Recognize volume as additive. Find volumes of solid figures composed of two nonoverlapping right rectangular prisms by adding the volumes of the nonoverlapping parts, applying this technique to solve real-world problems.	Investigate Graham Fletcher's website (www.gfletchy.com) to find 3 Act Tasks for understanding volume. *Packing Sugar* is a good one to start with.

3. What are some ways I can support my child's learning of measurement?

As you can see from the big ideas for measurement, measurement is a huge topic in elementary school math. As a parent, do the activities listed in the previous table with your child. You can help your child to see the relevance of measurement by drawing attention to the measuring you do each day. Also, keep in mind that measurement activities must be real-world applications, appropriately authentic for the child's level. My son Duncan isn't enormously curious about the distance from our home to the North Pole, but he is genuinely interested in the distance from our home to his school or to his favorite restaurant—especially since both are within bike-riding distance. Your child might not be too excited about the weight of a jumbo jet but might be eager to learn her birth weight and how much she's grown. While the capacity of an oil supertanker might amaze you (84 million gallons!), your child may be better able to connect to the concept of capacity by pouring cups of lemonade from a pitcher. So, talk about measurement in everyday contexts with your child. How far do you think it is to the market? How long will it take us to get to Grandma's house? How much heavier is the cantaloupe than the banana? Make measurement relevant and real world by making it part of your world—and your child's.

Also let your child help you with routine measurement tasks. For instance, involve your child in the kitchen. Unless you are an overconfident cook (like my husband) or a slightly lazy chef (like me), you'll be doing a lot of measuring. Home improvement often involves measurement—hopefully twice before you cut, right? Even doing laundry involves measuring—how many clothes to put in the washer; whether it is a small, medium, or large load (or super-large at our house); how much detergent to put in; and how long it will take before you should put the clothes in the dryer. Measuring is everywhere! Involve your child in these routines but also in the heavy-duty measurement tasks like ordering carpet or tile, buying fence materials, saving up for a big purchase, planning a road trip, and so on. When children see the usefulness in the mathematics they are learning at school, their motivation to learn increases and their math-positive mindset grows.

4. What do some common measurement terms mean?

The ability to communicate mathematically is viewed as a central goal for children (NCTM 2000). When children are familiar with mathematical terms, they can articulate their ideas with clarity and accuracy. When children are talking about measurement, for example they need to know that measurement is the comparison between an attribute of an object such as length or weight and a unit of measure for that attribute (Monroe 2006). The attribute being measured is the length and the unit might be inches or centimeters or chocolate chips, depending on the unit the child chooses. Knowing these terms and how to use them gives children confidence to talk about their mathematics experiences. Here are a few more measurement terms you might come across during math homework time, including kid- and parent-friendly definitions from the children's dictionary *The Easy, Simple, Fun Guide to Help Math Phobics Become Math Lovers* (Monroe 2006).

"PAUSE"ATIVE BOX

Common Measurement Terms

Take a moment to pause and review the following terms. Looking through the examples and discussing them together will help your child and you be on the same page when it comes to measurement terms.

Term	Definition	Example
Height	Refers to how tall someone or something is.	My height is five feet, five inches.
Length	Refers to the measure of the distance from one end of an object to the other. For a rectangle, the length usually refers to the longer of the two dimensions.	When we lay a person or object down, we measure its length instead of its height. (This might be fun to try: Measure your child standing up and laying down. Are they the same? Height and length are two ways of describing linear measurement.)
Width	The measure of an object from side to side. For a rectangle, the width usually refers to the shorter of the two dimensions.	The width of a standard football field is $53\frac{1}{3}$ yards.
Capacity	Refers to the amount of liquid a container can hold. However, it also might be used to refer to other types. (See examples.)	The capacity of the Cutler family minivan is 8 people. The capacity of the Cutler family maxivan is 12 people. The capacity of both vans for holding spilled soda is approximately 10 gallons or 37.85 liters.
Volume	Refers to how much space is occupied by a three-dimensional object. The volume of a space figure is how much space it occupies. The volume of a container is how much it can hold.	Volume is reported in cubic units. Each face of a cubic unit is a square and each edge is one unit in length. That's why volume is reported in cubic units. It's also why it's handy to fill rectangular prisms with little cubes to find volume before we introduce the formula *length* × *width* × *height*.

(continued)

(continued)

Term	Definition	Example
Mass	The amount of matter in an object. It is usually measured in grams and kilograms. The mass of an object remains the same regardless of location. Mass is measured by using a balance and comparing a known amount of matter to an unknown amount of matter.	Mass and weight are related. Mass measures matter in an object, and weight measures gravitational force on an object. For example, if I went to the moon, my mass would stay the same; my body would be the same shape and have the same number, type, and density of atoms. But my weight would change because the gravitational pull on the moon is much less than on earth (83.3% less).
Weight	Weight is how heavy an object is. It is the measure of the pull of gravity on the object. Weight changes depending on location. Your weight on the moon is different than it is on earth (while mass stays the same). Weight is measured on a scale.	Is a weight-positive mindset a thing? My weight on the moon is 23.1 pounds. I think I'll stick with that.
Area	Refers to the space or surface included within a set of lines. The area of a plane figure is the space enclosed by that figure. Area is also the measure of the enclosed space. Area is expressed in square units.	Area is a handy measurement for knowing how much carpet or tile to order. Measure the area twice, cut once.
Perimeter	The name for the outer edge of a figure or shape. It is also the distance around a shape or figure. *Circumference* is a special kind of perimeter—the distance around a circle.	Perimeter and area are related and sometimes children mix them up. A good way to remember the difference is to understand the prefix *peri* means to enclose or surround. Area, in common usage, means "a space."
Radius	Any line segment from the center of the circle to a point on the circle. The radius of a sphere is any line segment from the center of the sphere to a point on the sphere. Circles and spheres have an infinite number of radii (plural for radius).	The radius of the earth, since gravity causes it to bulge in the middle, is about 3,958 miles, give or take.

(continued)

(continued)

Term	Definition	Example
Diameter	The line that passes through the center of a circle and has endpoints on the circle. Since points have no length, area, or volume, there are an infinite number of points on a circle. There are, therefore, an infinite number of lines that can be drawn through the center of a circle to the edge. So, a circle (and a sphere) both have an infinite number of diameters. Cool, eh?	The diameter of the earth is about 7,917 miles. If you could dig a tunnel through the center of the earth, walking that distance would take about 39,585 hours or about $4\frac{1}{2}$ years without stopping.
Circumference	The distance around a circle, which equals a little more than three times the diameter of that circle.	The circumference of the earth is about 24,901 miles. A plane would be the way to go for that trip. The fastest, nonstop flight around the earth was undertaken in a B1 Lancer and completed in 36 hours, 13 minutes.
Customary System of Measurement	The measurement system used most commonly in the United States. The customary system was brought to the United States as the British Imperial Measurement System by colonists from England (who eventually adopted the metric system in 1965).	In elementary schools, the commonly used customary units include: • teaspoons, tablespoons, cups, pints, quarts, and gallons to measure dry and liquid volume; • inches, feet, and yards to measure capacity and volume (which is always measured in cubic units); • inches, feet, yards, and miles to measure length; • degrees Fahrenheit to measure temperature; and • ounces and pounds to measure weight.
Metric System	A decimal-based system of measurement used in almost every industrialized country in the world except the United States, which maintains the customary system for most areas except the sciences. The metric system is handy partly because it includes a standard set of interrelated base units (meters and grams, for example) and a standard set of prefixes in powers of ten (like centi- and milli-). These base units are used to generate larger and smaller units.	Metric base units include grams and meters as well as Celsius for measuring temperature. The metric system is the international and scientific standard used around the world.

TEACHING TIP

Teaching Measurement Terms

Here are several ideas for building children's measurement vocabulary in meaningful ways:

Math Word Wall

Consider displaying a Math Word Wall if you do not already take advantage of this interactive tool. High-frequency, unit-specific, and/or problematic terms are displayed where students can refer to them for their discussions, writing, and problem comprehension. Students can help you choose words from the curriculum and help you create a display that is useful to them. Distinguish easily confused terms such as *perimeter* and *area* by writing them on different colors of paper. Use graphics and drawings rather than text-heavy definitions to help students "see" the meaning of mathematical terms. Review the terms often with frequent interaction with the Math Word Wall.

Mind Reader

Mind Reader is a game where students use deductive logic to identify a term on the Math Word Wall. To play, choose a "mystery word" from the Word Wall. Give several clues to help students identify the word. The first clue should be general and apply to several terms on the Word Wall. Each subsequent clue should get more specific to help students to narrow down the words on the Word Wall. As you give each clue, students should write down all the words that match the clues given so far and cross off words that no longer apply. The last clue should be the most obvious one. For example, for the word *circumference*, the clues might be:

1. The mystery word has something to do with shapes.
2. The mystery word has something to do with circles.
3. The mystery word has something to do with measurement.
4. The mystery word is a special kind of perimeter.
5. The mystery word means the distance around the edge of a circle.

(continued)

(continued)

Flyswatter

This is another fun activity for reviewing mathematical terms. Give two students flyswatters and call out definitions or examples that relate to a word displayed on the Word Wall. Whoever swats the word first is the winner.

You might also have students occasionally rearrange the Word Wall, clustering related terms or using string to connect terms to form a web. Remember that you don't have to display all the mathematical vocabulary for your grade level at the same time. Students tend to ignore an overloaded Word Wall. Instead, introduce terms as you begin a unit and rotate them out as students master them. In whatever way you structure your Math Word Wall, if students are using it to improve their comprehension and communication in mathematics, it is worth the effort.

An example of a Math Word Wall from my classroom.

5. My children understand better when they try things out for themselves. What are some materials to help children understand measurement?

Math-positive parents recognize that learning things by doing allows children to construct their own understanding rather than being passive receivers of information that is easily misunderstood or forgotten. You might remember an elementary school math worksheet with a picture of a nail positioned so it's lying above a segment of a ruler. The nail is perfectly aligned with the left end of the ruler and the child need only read the number where the nail ends to determine the nail's length. This is a common format for measurement worksheets, but why waste a tree on a worksheet that does little to build a concept when you can do something active? Measurement makes the most sense to children when they can use materials to explore, estimate, experiment, and experience. When children choose *what* they measure and what they *measure with*, their investment in the measurement process is personal and meaningful. To help a child build understanding of measurement concepts, a classroom (or a math-focused home) might be stocked with a variety of items like those found in the table that follows.

"PAUSE"ATIVE BOX

Manipulatives for Use with Measurement

Here is a list of manipulatives most frequently stocked in classrooms for supporting the learning of measurement. Take a moment to pause and review the list. I've also given a suggestion for a simple measurement task using each type of manipulative. Put an asterisk by the activities you'd like to try with your child. Try them out—how did it go? Have fun building a math-positive mindset along with your child.

Manipulative	Description	Uses in Measurement	Math Activities
Nonstandard Measurement			
Counters	Teddy bears, dinosaurs, inchworms, Unifix or Snap Cubes, frogs, or any other type of counter can be used as a nonstandard unit of measurement.	measure length, weight, and/or capacity in nonstandard units A fork might be 7 cubes long.	• To measure length, have children set counters end-to-end with no gaps beside an object. The number of counters gives you the length. For example, a fork might be 7 cubes long. • To measure weight, have children set an object on one side of a pan balance then place counters on the opposite side of the balance. For example, a spoon might weigh the same as 22 cubes. • To measure capacity, have children fill a container with counters. For example, a coffee mug might hold 37 cubes.
Color Tiles and Cubes	Color tiles are one-inch square tiles in four colors (usually red, green, blue, and yellow). Color cubes are one-inch cubes that come in lots of colors.	measuring perimeter, area, length, capacity, and weight useful for transitioning children from nonstandard to standard units since they approximate the standard unit	Color tiles are great for measuring the area of a square. Work with children to trim a sheet of paper to measure 8 square inches. Have children count the total number of tiles used to cover the paper to find the area of the paper in color tiles.

(continued)

(continued)

Manipulative	Description	Uses in Measurement	Math Activities
Open Materials	Open materials are items that can be used in multiple ways. They include pipe cleaners, cotton swabs, beans, dry cereal, drinking straws, paper clips, flat-sided marbles, craft sticks, and so on.	measuring in nonstandard units	Have children use cotton balls to measure length by laying the cotton balls end-to-end; to measure capacity by filling a jar with them; or to measure area by covering a shape with them.
Pan Balance	Sometimes called a balance scale or a bucket balance, a pan balance has a beam with two cups or pans on either end. When the cups contain the same weight, the beam is balanced.	measure weight using simple comparison ("Is a glue stick lighter than a bottle of glue?"), nonstandard units ("How many teddy bears does a glue bottle weigh?"), or standard units ("How many grams does a glue bottle weigh?")	Give children time to explore using a pan balance. Have them place an object in one cup and use standard units (such as grams) or nonstandard units (such as teddy bears) to discover the weight of the object. Or they can use the balance to discover which object weighs more—the side that dips holds the heavier object.
Standard Measurement			
Linear Measurement Tools	centimeter cubes, inch cubes, ruler, yard stick, meter stick, trundle wheel, and measuring tape	measuring in standard units	Children are ready for standard units in second grade. Connect their previous experience with nonstandard units that approximate standard units (like centimeter cubes and inch tiles) to make this jump less drastic.
Angle Measurement Tools	compass, angle ruler, protractor	measuring angles in standard units	Children begin using a protractor to measure angles in about fourth grade, depending on the state objectives. Try this: Have children write their names using block letters then measure the angles found in each letter. For example, the letter E has four right angles. Cool!

(continued)

(continued)

Manipulative	Description	Uses in Measurement	Math Activities
Weight Measurement Tools	metric and customary weight (or mass) sets, pan balance, and digital or platform scales	measuring weight in standard units	A kitchen scale is all you need to explore standard units for weight. Have children explore and compare the weights of fruits and vegetables.
Capacity Measurement Tools	metric and customary measuring cups and spoons, graduated cylinders, beakers, and measuring jars	measuring capacity in standard units	The kitchen or cooking center is the perfect spot for measuring capacity. How accurate are the measuring cups in your cupboards? Have children compare several labeled with identical capacities to find out.
Temperature Measurement Tools	metric and customary thermometers	measuring temperature in standard units	Children can practice measuring temperature and making science connections by placing two thermometers outside—one in a sunny spot and one in the shade. Read them every few hours to compare the differences.
Time Measurement Tools	digital and analog clocks, stopwatch, timer, and geared clock (Geared clocks have hands that are synchronized to move in conjunction with one another. They help children see the relationship between the minute and hour hand because as the minute hand makes one complete revolution in 60 minutes, the hour hand moves to the next hour.)	measuring time in standard units	What clocks are in your home? Do you refer to the microwave's digital display or simply ask a robot assistant for the time? Today's children don't have much practice reading analog clocks. To help with this, consider hanging an analog clock next to a digital one in your home. Children can compare the two devices as they learn this timeless skill.

6. Why is estimation an important part of the measurement process?

We live in a world of exactness. Milk is sold in pints, quarts, half-gallons, and gallons. Watt-hours measure our electricity consumption. Police don't just guess miles per hour when catching speeders on the highway. So why is estimation an important part of the measuring experience? The act of making an estimate before measuring helps to build a child's number sense. Revising an estimate partway through the measurement process also helps children see how reasoning, rather than whimsy, confirms or rejects our estimates. Recently, my son Duncan and I had a measurement conversation that went something like this:

Mom: Duncan, I wonder if you have enough time to finish practicing your piano lesson, straightening your room, and weeding the garden before we must leave for flag football?

Duncan: I can do it, Mom. When do we have to leave?

Mom: In forty-five minutes. I wonder how long each of those jobs will take?

Duncan: If I have three jobs and forty-five minutes, that's fifteen minutes for each job.

Mom: Sounds good. Give it your best shot.

(Duncan leaves and returns twenty-five minutes later.)

Mom: Have you finished everything already?

Duncan: (looking at the clock) I thought I could do it all in forty-five minutes, but I only have my bedroom straightened, and it's been almost thirty minutes. I should have two jobs done.

Mom: Hmm . . . what are you thinking?

Duncan: I'm thinking of asking my brothers to help me weed the garden. With all of us working together, we can have it done in five minutes. That way I can practice my piano lesson and be ready to go in time for football.

Mom: Good problem solving. Let me know how it goes.

Duncan got it! He incorporated estimation into measurement in a practical way that helped him solve a measurement problem. Include estimation next time your children have a measurement problem and watch their number sense and math-positive mindsets grow!

Ideas for Encouraging Estimation

TEACHING TIP

Estimation is a key element of number sense. Students' reasoning skills are supported when they are given frequent opportunities to estimate. Note that a one-off guess is *not* a reason-based estimate and does little to build students' sense of number. Following are ideas for supporting students' estimation skills.

Length Activities

For young children, length activities can help build estimation skills. Kindergartners, for example, might be asked to estimate, then measure the length of their desk using Unifix cubes. Their initial estimate will probably be a shot in the dark, perhaps their favorite number or the number 100 (a rock 'n' roll number for kindergartners). Allow students to get about halfway through laying the cubes along the edge of the desk. Then ask them to revisit their estimate. Do they still think it will take 100 cubes? Revising an estimate based on more information is not "cheating." It's using reasoning to make sense of number.

Estimation Learning Stations

With older students, you can encourage reasoning during estimation by having an estimation learning station for capacity. Each day, students can use the information from the previous

(continued)

(continued)

days' exploration to build reasoning. On Day 1, students count out thirty toy cars used to fill a glass jar. On Day 2, the jar is only halfway filled. How many cars does it contain? On Day 3, the jar is filled two-thirds of the way. Students use the benchmark of thirty for a full jar as a reference to reason about the number of cars in the jar on the following days. Changing the contents from small items to large items helps build an understanding of the inverse relationship between the size of the item and the number needed to fill the jar. The smaller the item, the more you need to fill the jar. The larger the item, the fewer you'll need. This proportional reasoning comes into play in many areas of mathematical reasoning, estimation, and number sense.

7. What is a measurement benchmark? How does using benchmarks help children to measure more accurately?

A benchmark is a quick reminder of the size of a unit. For standard unit measurement, real-world benchmarks can help build measurement sense. For example, the height of a door can be estimated using a benchmark of the height of a six-foot-tall adult. Or the length of a large paper clip is about one inch.

"PAUSE"ATIVE BOX

Using Measurement Benchmarks

If you would like to help your child discover how fingers can approximate some standard units, pause and do the following activity together.

1. Ask, "How long is an inch?" Then give the task, "If your ruler has units marked on each side, find the side of the ruler with larger spaces between the numbers."
2. Confirm and show, "This is the side of the ruler that measures inches. The distance between the whole numbers on the ruler is always the same, right? This is one inch."
3. Now propose the scenario, "What if you didn't have a ruler handy but still wanted to be able to measure in inches? What could you do?" Your child may suggest that you use your fingers to measure about an inch.
4. Say, "Look at your index finger. Hold the ruler against your finger. Is the knuckle of the finger to the tip of the finger about one inch long?"
5. Continue, "Now look at your thumb. Is the tip of your thumb to the knuckle about one inch long? Are any other sections of your fingers about one inch? Try checking with the ruler."
6. Ask, "How long is a foot?"
7. Confirm, "The distance from one end of the ruler to the other end is one foot."
8. Have your child hold a ruler against their arm. Ask, "Is the length from your elbow to your fingertip about a foot long? How about from your ankle to your knee?"
9. Finally give your child time to explore, "Try using your new benchmarks for one inch and one foot to measure some objects. Then check with the ruler to see how close your informal measuring is to the actual measurement."

8. What is data analysis and why is it taught in elementary school?

Our society is inundated with data. We live in an increasingly complex world with information constantly thrust upon us by a pinging computer we carry in our pocket. Some of that information is useful; however, some of that information is also deceptive or based on misinformation, hidden agendas, and so forth. For example, a box of triple fudge brownies says that each brownie contains "Just 100 Calories." Sounds fabulous, right? However, at a closer look, the serving size of that 100-calorie triple fudge brownie is actually a one-half inch square, not the typical one-inch square found in the regular brownie package. Now that's just mean!

A literate, functional society is dependent on citizens who can think critically—who can analyze and interpret data, study options, and make informed decisions. It's not enough to teach children to blindly follow mathematical procedures; they must analyze and reason to evaluate the credibility of information and arrive at their own conclusions.

Data analysis connects ideas and procedures from number, geometry, measurement, and algebra. It offers a natural way for students to bridge mathematics with other school subjects and with experiences in their daily lives. Giving children experience asking questions, questioning inferences drawn from incomplete or biased data, gathering data, analyzing it, and interpreting results—all within a context they care about—ensures they are prepared to apply these skills in the future. When they buy a car, vote for a candidate, or consider the pros and cons of a career, they are putting to use essential data analysis skills practiced in their classroom and (hopefully math-focused) home.

9. I'm used to graphs and charts in a business setting, but what does data analysis look like in elementary school?

Data analysis sounds like something from a forensic crime show or a scientific experiment. But data analysis is part of mathematics (and the real world) from an early age. Initial experiences with data analysis involve children's own environments—their preferences and experiences.

Data collecting starts simply and consists of real-world objects. A kindergarten class might compare the different shoes they have—lace-ups versus slip-on versus Velcro. Representing data starts simply, too. The same kindergarten class might then create a pictograph (showing pictures of the real-world objects) or a bar graph.

Shoes We Wear

Example pictograph.

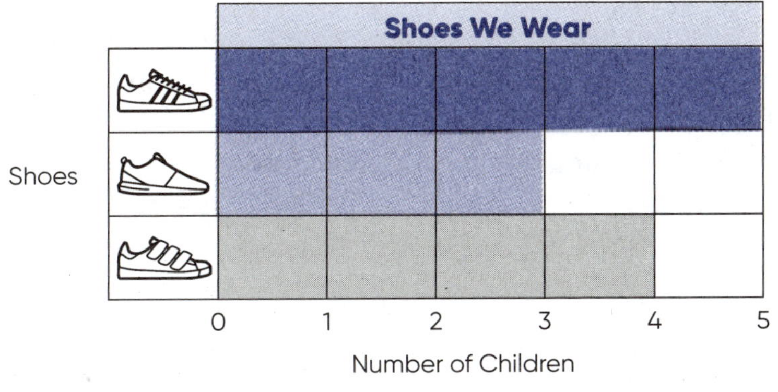

Example bar graph.

Initially, children will sort data into two categories. For example: *Do you like vegetables? Yes or No.* Later on, they might be offered several choices. *Do you prefer carrots, green beans, or lima beans?* The bounds of the data collection continue to expand with open response questions: *Which is your favorite vegetable?*

To analyze the data they have collected, younger children might compare the votes of their classmates. Older children might survey children from another class to find out their preferences for vegetables.

To present their data, the younger children might create a graph they hang in the classroom. The older children might pass the results of their multi-class survey on to the cafeteria supervisor, thus showing how data collection and analysis helps us to understand and solve problems.

Just like the reach of the survey expands as children grow older, graphs become more varied as children become more capable of reasoning abstractly. Pictographs and bar graphs are introduced first with one bar representing one unit (one vote for Brussel sprouts). In about third grade, children begin using scaled bar graphs with one bar representing multiple units (ten votes for Brussel sprouts).

It's not enough just to gather and represent data; we must also interpret and try to understand what the information means. When young children talk about graphs, conversations center on simple comparisons—*how many more children like Brussel sprouts better than lima beans?* Older children formulate hypotheses about data; they might look at the graph and hypothesize that *a new student in the class is likely to also love Brussel sprouts because they're the indisputable favorite in the vegetable survey.*

10. What are some real-world examples of data analysis that I can show to children?

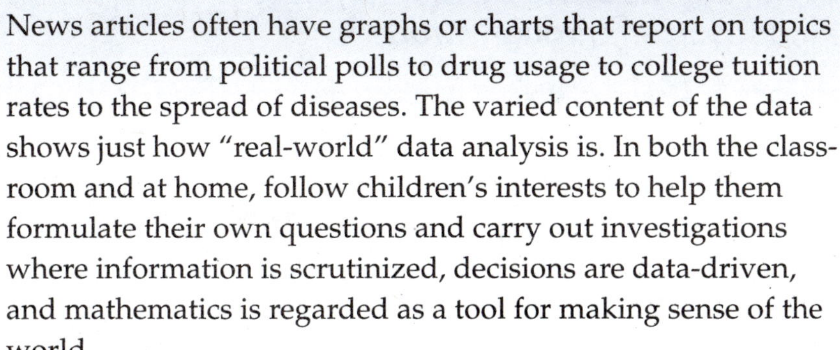

News articles often have graphs or charts that report on topics that range from political polls to drug usage to college tuition rates to the spread of diseases. The varied content of the data shows just how "real-world" data analysis is. In both the classroom and at home, follow children's interests to help them formulate their own questions and carry out investigations where information is scrutinized, decisions are data-driven, and mathematics is regarded as a tool for making sense of the world.

"PAUSE"ATIVE BOX

Real-World Data Activity

Take a moment to pause and find a graph or chart. News articles, magazines, and so forth are great places to start. Point the graph out to your child and initiate a conversation. You might discuss whether the graph shows categorical or numerical data. If it's numerical, is it discrete or continuous? (See Question No. 12 in this chapter for further insight into these terms.) Is the scale accurate or does it skew the data? Finally, be sure that you ask questions about how the data may be used. Will the information change opinions about something? Will it help people plan for the future? Do the data suggest a trend?

11. What are the big ideas for data analysis in elementary school?

Most data children work with should focus on themselves and their own lives. Surveys about their favorite things (colors, books, foods, places to visit), numbers they can report (siblings, pets, or buttons on their clothes), and other kid-focused ideas make data analysis relevant and fun. See the "Pause"ative Box in this section for a summary of the big ideas.

"PAUSE"ATIVE BOX

Summary of Big Ideas for Data Analysis for Kindergarten through Grade 5

Here are some of the big ideas related to data analysis children will learn in school, including insight into what might be happening in their classroom (National Governors Association Center for Best Practices and Council of Chief State School Officers 2010). Pause and take a moment to review the list. How might you modify the classroom examples for use at home? It might be hard to conduct a survey at home (unless you have a large family), but this table offers suggestions anyone can try out simply by opening the kitchen cupboard. Give these ideas a try and watch your children's math-positive mindsets about data analysis grow!

Grade	Big Ideas for Data Analysis	Classroom Examples	Math Activities
K–2	Generate questions, collect and organize data, present data to answer questions.	The teacher asks, "I wonder how many kids in our class have pets? Raise your hand if you have a dog. Who has a cat? A fish? A gerbil? A giraffe?" Children then use real objects, pictures, and graphs to represent data and make simple comparisons between sets of data: "Let's use stuffed animals to represent the pets owned by the kids in our class. Lay all the dogs in a line. Then line up the cats. We are building a bar graph with objects that represent our pets. Which type of pet is the most popular? How many more cats are there than giraffes?"	Have children clean out the pantry or food storage area in their house and sort the cans—fruits, vegetables, soups, and so on. Children arrange the cans in rows then make comparisons. (This is a concrete bar graph, meaning it is made of real objects instead of pictures.) Use terms like *most*, *least*, *more than*, *less than*, and *equal to* in your discussion.

(continued)

(continued)

Grade	Big Ideas for Data Analysis	Classroom Examples	Math Activities
3–5	Consider how the method of data collection (such as surveys, experiments, and observations) might affect the results.	The teacher asks, "I wonder what the most popular condiment is on the salad bar in the school cafeteria? We could do a survey and ask people what they like on their salad, or we could sit near the salad bar and watch what people choose." "I wonder if both methods would give us the same results. Would people report that they prefer fat-free dressing but choose full-fat dressing at the salad bar?"	Talk with children about which type of food you have the most of and why that might be. Do you buy more fruit and that's why it is most commonly found in your kitchen or does it not get eaten as quickly as soup? Would checking the expiration dates provide more data? Would watching the family eat dinner for a week provide additional data?
	Understand the differences in representing categorical data and numerical data.	Categorical data in the previous example would be the different types of dressing to choose from; numerical data would be the amount of dressing poured onto the salad.	Categorical data in the pantry example would be the different types of foods. Numerical data would be the amount of food in the cans. Taking an inventory of the school cafeteria shelves may be fascinating! To learn more about categorical and numerical data, see Question No. 12 in this chapter.
	Organize data in a table; represent the information using line plots, bar graphs, and/or line graphs.	Students explain, "Our bar graph shows that full-fat dressing was chosen more often than fat-free dressing. Our line graph shows that between 11:30 and 11:40 thirteen people chose fat-free dressing, between 11:40 and 11:50 six people chose fat-free dressing, and between 11:50 and 12:00 two people chose fat-free dressing."	Ask children how they could transfer the information from the concrete bar graph to a different graph. They might choose to draw a pictograph, create a bar graph, or use a table to record the number of cans in each category.

(continued)

(continued)

Grade	Big Ideas for Data Analysis	Classroom Examples	Math Activities
3–5	Increase analysis of the data gathered and represented; move beyond simple visual comparisons of graphs to focus on how the data are distributed.	Students question, "I wonder why people who eat lunch later choose full-fat dressing more often than people who eat lunch earlier in the day? Could it be that hungrier people care less about calories?"	The data gathered from the pantry graphing activity could be used to help write the next week's grocery shopping list. What do we need the most of? What can we skip buying this week?
	Use measures of center (median, mean, and mode) and understand what each does and does not reveal about the data set.	Students share, "When we look at preferences for dressing, we see that the mean is fat-free ranch, the median is full-fat ranch, and the mode is fat-free ranch. That tells us that the cafeteria employees might want to order plenty of fat-free ranch but not forget to order a bit of full-fat ranch, too."	Imagine that the pantry graph showed 8 cans of soup, 10 cans of fruit, and 15 cans of vegetables. Talk with children about which measure of central tendency (median, mean, or mode) is most useful in planning a grocery trip. • The *mean* would tell us the average number of each type of food in the pantry. We add up all the cans then divide by the number of categories. $8 + 10 + 15 = 33$ $33 \div 3 = 11$ Should we buy 3 cans of soup and 1 can of fruit to make it "even"? • The *median* describes the middle number in the data set. If we list the number of cans in each category in order from smallest to largest—8, 10, 15—the median would be 10. Would we buy 2 cans of soup to make it reach the median? • The *mode* is the number that occurs most often in a set of data. In our pantry example, we have no mode because no number appears more often than any other.

12. Just when my children seem to get the hang of one graph, they must do it a different way. Why do we need these different types of graphs? Can't a bar graph be used for every type of data?

Because there are different types of data, we need different types of graphs to represent the data accurately. There are two main types of data with which elementary-age children work: categorical data and numerical data. Let's take a closer look at each.

Categorical Data

This gives us data in the form of words. The Favorite Pets graphs shown in Question No. 13 are examples of categorical data. We created bar graphs to display results of voting for favorite pets. The data were categorical because the types of pets do not themselves have numerical value. Examples of categorical data elementary-age children might investigate include favorites (movies, books, colors, television shows, ice cream, cars, celebrities, etc.), physical characteristics (eye or hair color, color of clothing, etc.), and other topics that children encounter on a regular basis.

Numerical Data

This gives us data in the form of numbers or measurements. What if we wanted to do another survey to find out how many pets nine classmates have? We might gather the following numerical data: 0, 0, 1, 1, 1, 2, 2, 4, 11. The ninth friend might have a cat that just had kittens. Yikes! These data are numerical because, well, they're numbers. Numerical data can be further broken down to *discrete* and *continuous*. Discrete data can be counted. So, our example of how many pets nine classmates have is discrete data. We can count those pets. But what if we asked about the weight gain of those eleven adorable kittens? That can only be measured, giving us continuous data that moves along a continuum. In elementary school, we represent continuous data on a scatter plot or line graph. See the following graph for an example of continuous data representing the weight gain of the kittens.

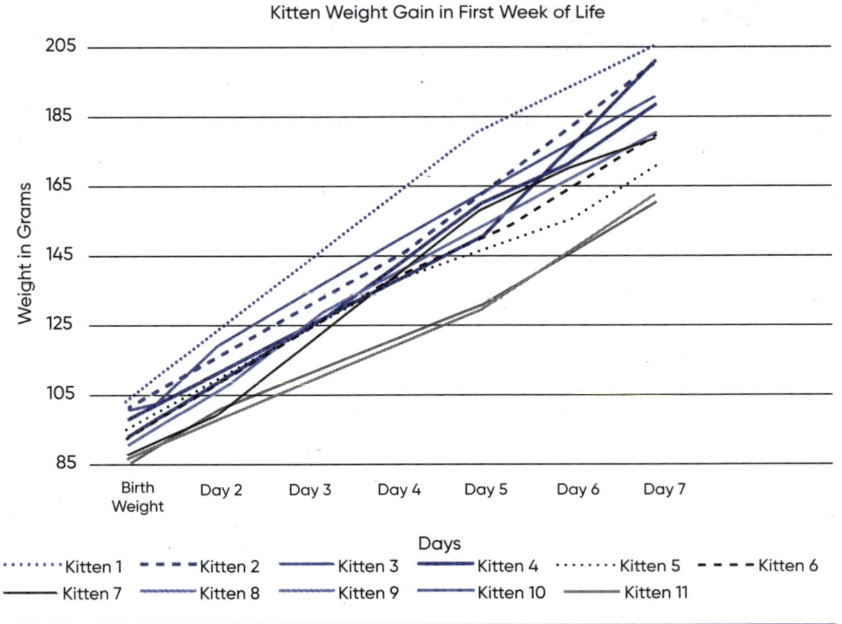

Continuous data representing the weight gain of kittens.

The line graph is handy for displaying continuous data since those measures fall in a range but cannot be counted or separated into distinct values. They must be thought of as intervals on the real number line (think fractions and decimals). For example, the kittens' growth can be measured in grams, tenths of a gram (decigram), hundredths of a gram (centigram) or thousandths of a gram (milligram), or even a quintillionth of a gram (attogram—cool!). There is always a smaller, more accurate measurement we could make. In this way, continuous data can be thought of as being uncountable—infinite. We usually round the number off to make it less complicated for gathering, reporting, and analyzing data.

"PAUSE"ATIVE BOX

Types of Graphs

Here are some graphs typically taught in elementary school. Pause and take a moment to review the list. Notice how some graphs are used for categorical data and some are used for numerical data. With which graphs would you like to become more familiar?

Graph	Example(s)
Categorical Data	
Real Object Graph	
Picture Graph (sometimes called a pictograph)	
Bar Graph	

(continued)

(continued)

Graph	Example(s)

Numerical Data

Frequency Table

(sometimes called a frequency distribution table)

How Many Children Families in My Neighborhood Have

Number of Children	Tally Marks	Frequency (Number of Families)
0	|||	3
1	||	2
2	|||| |	6
3	|||	3
4	|	1

How Many Children Families in My Neighborhood Have

Number of Children	Frequency (Number of Families)
0	3
1	2
2	6
3	3
4	1

Circle Graph

(sometimes called a pie graph)

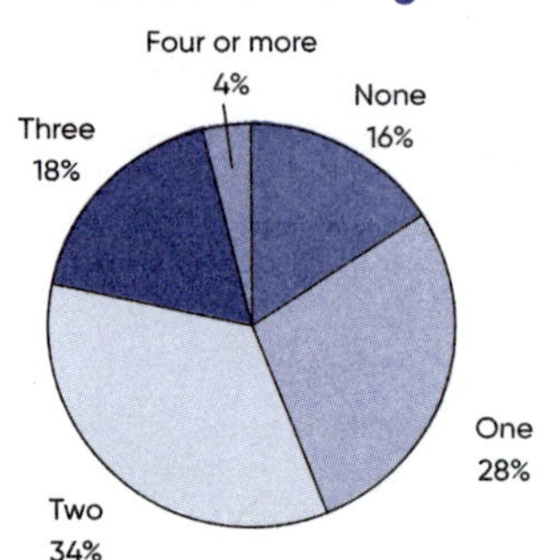

Dot Plot

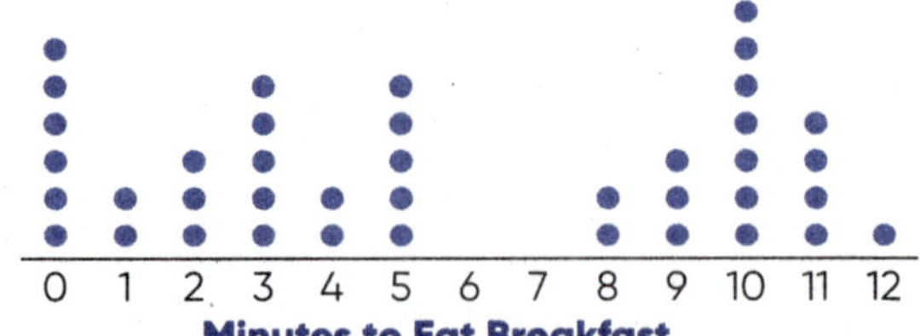

Histogram

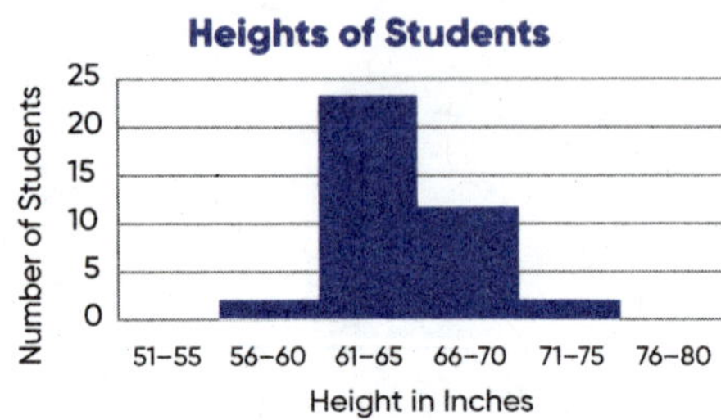

(continued)

(continued)

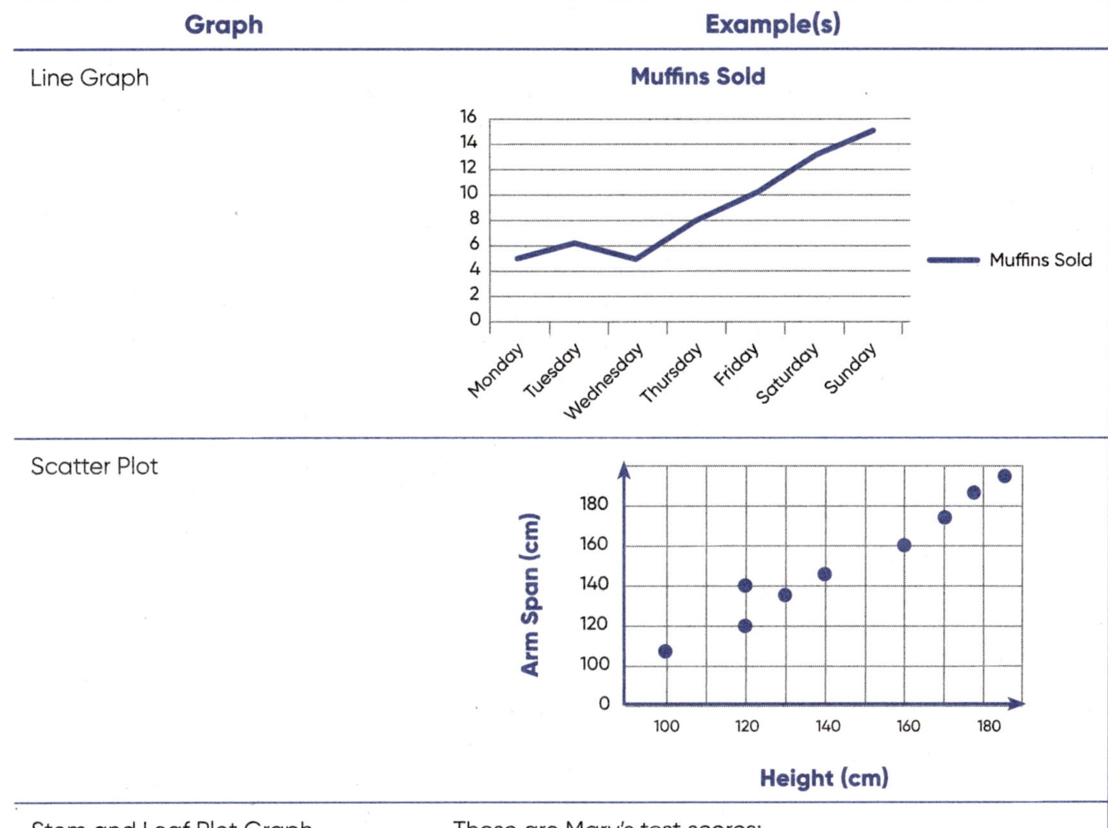

Graph	Example(s)
Stem and Leaf Plot Graph	These are Mary's test scores: 72, 49, 62, 58, 73, 55, 78, 83, 57, 63, 73, 73, 75, 85, 85, 64, 61, 67, 75, 91 The stem-and-leaf plot for her scores is shown below.

Stem	Leaf
4	9
5	5 7 8
6	1 2 3 4 7
7	2 3 3 3 5 5 8
8	3 5 5
9	1

TEACHING TIP

Graph of the Day Activity

Graphs can be used regularly in the classroom to start mathematical conversations. Some teachers use a *Graph of the Day* activity as a warm-up. This has several purposes. It may be used to take the attendance, to give children a chance to express their opinions, or to get some insight into children's worlds. You might laminate a poster board to use and reuse for these graphs. On one side of the poster, make two columns. On the other side, make three columns. Leave a bit of space at the top for writing or posting the question of the day and label the columns with the available responses using a dry erase marker. For the question, "Do you prefer books about cars, animals, or people?" label the columns *Cars*, *Animals*, and *People*. Children respond by writing their name, hanging their name card, or attaching a photo of themselves in the appropriate column. Spending a few minutes discussing and interpreting the data builds mathematical vocabulary, oral language, and an understanding of how data is used to make decisions. Explain to the class that you intend to use this information the next time you are ordering books for the classroom library. This models for children how the data analysis process works to inform our decisions. Here are a few more ideas for a graph of the day:

- Two-column graphs work well for many questions, including any Yes or No question. "Did you bring lunch from home?" "Did you brush your teeth this morning?" "Do your shoes have laces?" "Are you wearing snow boots today?" "Do you have your phone number memorized?"
- Three-column graphs work well for limiting children's options when too many choices would be overwhelming. For example, "Do you like chocolate chip, oatmeal raisin, or snickerdoodle cookies?" works better than "What is your favorite type of cookie?" Other examples could be, "Would you rather visit Hawaii, Alaska, or New York?" or "Would you rather own a pet lizard, ferret, or snake?"

13. What does *scale* refer to in data analysis?

On most graphs, a scale is a series of marks at regular intervals along the side. While the scale could count by any number, graphs for young children usually maintain a one-to-one correspondence, meaning one vote for broccoli gets one colored-in bar on the broccoli column of the bar graph. Later on, the scale can be modified to allow for larger units to be represented. One bar might stand for 20 votes for broccoli. Or on a pictograph each picture of a lima bean might stand for 3 votes for lima beans. Scale can be manipulated to skew data. Starting in about third grade, children are asked to compare different representations of the same data and evaluate how well each representation shows important aspects of the data. Take, for example, the following graphs, created after collecting data about favorite pets.

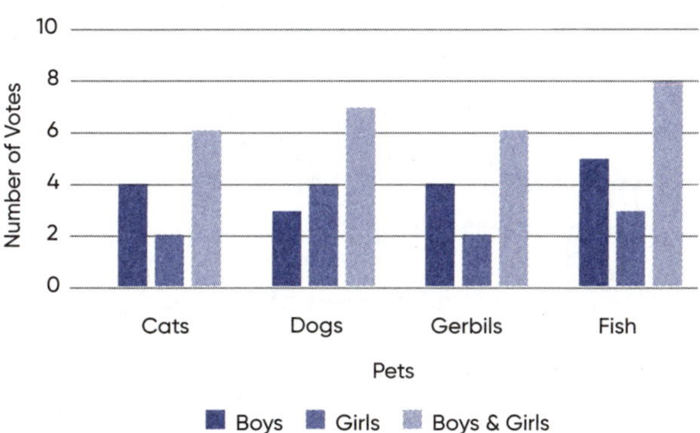

A graph about pets shows that fish are the favorite when we combine the votes of boys and girls.

Though both graphs display the same data, it is easier to make comparisons between votes for favorite pets when the graph is scaled as in Graph A on the previous page. In Graph B, the scale of the y-axis (the number of votes) starts at 2, which omits the girls' votes for cats and gerbils. Sliding the scale creates a misleading graph.

Graph B: Favorite Pet

Number of Votes
17
12
7
2
Cats Dogs Gerbils Fish
Pets
Boys Girls Boys & Girls

Sliding the scale on the Favorite Pet graph makes it difficult to analyze the votes and leaves out cat and gerbil lovers entirely!

14. What are some materials for helping children understand data analysis?

Unlike the other content areas in mathematics, data analysis does not require specific tools or manipulatives. Though you may wish to supply graph paper to make drawing tables and graphs more precise, all you really need is an interest in using data to guide decisions and solve problems. You can engage children in relevant, real-world data analysis experiences by simply opening the sock drawer or kitchen cabinet. Follow through the sequence of collecting, organizing, representing, and interpreting data to answer questions that are important in your math-focused classroom or home.

QUICK REFERENCE CHART

Common Concerns about Measurement and Data

This table is a quick reference for some of the *most common concerns* I hear from parents. The table lists the concern, an explanation of the thinking behind it, and an idea that addresses the concern. Consider photocopying this table and placing it somewhere where you can be reminded of what is needed to build your math-positive mindset and that of your children and/or students.

Concern	Explanation	What to Do	Question Number(s) (to Learn More)
Why measurement?	Much of our real-world mathematics involves measurement. Hitting the snooze button, measuring oatmeal, filling up the gas tank, paying the cashier, driving to work, checking the clock, speeding up or slowing down, punching the clock—all measurement concepts that fill our math-positive day.	The clock, the gas tank, the bank account, the speedometer, the workday are all conversation starters that illustrate the real-world nature of measurement. Your child need only hear about the first few hours of your day to know that learning measurement in a valuable real-world skill.	1
What are measurement concepts for elementary school?	Time, temperature, money, length, area, volume, and weight are all measurable attributes children learn about in elementary school.	Support measurement learning by having young children find objects around the house that are longer than/shorter than their shoes. Your third, fourth, and fifth graders can find things that are exactly the length of a ruler (one foot).	1, 2
Supporting measurement understanding at home	When children see practical applications for the mathematics they are learning at school, their motivation to learn increases and their math-positive mindset grows.	Involve your child in daily measurement routines like cooking but also in the heavy-duty measurement tasks like ordering carpet or tile, buying fence materials, planning a road trip, and so on.	2, 3

(continued)

(continued)

Concern	Explanation	What to Do	Question Number(s) (to Learn More)
Where can I find child-friendly definitions of measurement terms?	Measurement terms like *length, height,* and *width* are easy because they are used often, but less-familiar terms like *perimeter, area, circumference,* and *volume* can be tricky for children to understand.	A handy child-friendly math dictionary can provide simple definitions of measurement terms and illustrations. Check out *Math Dictionary: The Easy, Simple, Fun Guide to Help Math Phobics Become Math Lovers* (Monroe 2006).	4
What measurement tools should I have on hand?	Measurement must be taught using hands-on tools to solve problems situated in real-world scenarios. When children use measurement tools, they see that doing mathematics is active rather than passive.	Math-positive parents recognize that learning things by doing allows children to construct their own understanding rather than being passive receivers of information that is easily misunderstood or forgotten. Some handy tools for measurement include: pan balance, ruler, thermometer, clock, stop watch, and measuring cups. Older kids might like to use a compass to draw circles and a protractor to measure angles.	5
Estimation and measurement	Estimation becomes a tool for building number sense when we pause part-way through the measuring process to check the reasonableness of our estimate.	Show your child a small cup filled with dry beans. Ask them to estimate the number in the cup. Dump out the cup and count the beans, placing them back in the cup until it is half full. Have your child revise their estimate based on this new information. Finish counting out the remaining beans to find the capacity of the cup.	6

(continued)

(continued)

Concern	Explanation	What to Do	Question Number(s) (to Learn More)
		Next time, use the same cup but fill it with something larger like mini marshmallows. Your child should use number sense based on the bean experience to make an estimate. Repeat this process regularly to build ideas of capacity, measurement, and number sense.	
Benchmarks in measurement	Benchmarks in measurement are reference points for commonly used units like feet, inches, and pounds.	Have your child find a part of their body that is about one centimeter, about one inch, and about one foot. These benchmarks are reference points that can be used to check the reasonableness of answers to measurement problems or to make estimates.	7
Why data analysis?	Our society is inundated with data. Some of that information is useful but some is misleading. Today's children need to be critical consumers of data.	Go on a news article hunt for surveys, polls, graphs, and tables. The weather page always contains a few interesting snippets of data. How can this data help your family plan their day? Are there ways the data could be misleading? Talk with children about how data helps us make informed decisions.	8
What are the big ideas for data analysis in elementary school?	Children collect data from their own experiences, represent it in graph form, and interpret the data to try to understand what the information means.	At home, children can make a graph of the types of canned goods in the kitchen. What does this graph tell us? We love fruits but not vegetables? We don't eat soup very quickly? Data like this can help in planning the next grocery trip.	8, 9, 11

(continued)

(continued)

Concern	Explanation	What to Do	Question Number(s) (to Learn More)
Real-world data analysis	Businesses, schools, doctors, marketing teams, and government leaders all use data analysis to guide decisions.	Find examples of graphs in news articles and magazines. Talk about the graphs with children. Ask questions like: What does the data show? Why might the data be helpful? Is anything about the graph misleading, such as the scale? Discuss how data analysis helps us gather, organize, represent, and interpret information that is useful for making decisions.	10, 13
Types of data and types of graphs	*Categorical data* comes in the form of words (for example, preference for chocolate or caramel ice cream topping) and can be represented with a bar graph. *Numerical data* can be *discrete* (number of spoonfuls of chocolate sauce) and represented with a line plot or *continuous* (grams of chocolate sauce) and represented with a line graph.	Different types of data need different representations so that we can compare, interpret, and hypothesize. Try this activity with children: pour coins out of a piggy bank. Sort the coins into groups. Together make a bar graph using the coins. Ask questions: Which coin is most common? Which coin is least common? To experience numerical data, together make a line graph showing the total value of coins in the piggy bank. Every week, have children recount the coins to see how the value increases or decreases.	12, 13
Data analysis tools	You don't need fancy math manipulatives to build understanding of data analysis. Everyday objects from canned goods to socks and shoes make simple tools for investigations.	Use the contents of the kitchen "junk drawer" to build a real object graph. First, sort the items by type then lay them out in columns. Talk together about the graph. "Are there more writing items or cutting items? How many elastics are there? Why do you think there are so many batteries?"	14

I Can Help with Algebra

CHAPTER 7

TEACHING TIPS

1. I thought algebra was a high school math class. Why is my child expected to learn algebra at such a young age?

Yes, we all had an algebra class in high school, but we now know children need to understand algebra concepts early so they're ready for that *x*, *y*, *z* stuff later on. Algebraic reasoning is all about identifying relationships and generalizing those ideas to new situations or objects (NCTM 2000). Think about your day and how much of it is spent effortlessly following patterns you've established. Waking, bathing, dressing—even brushing your teeth is probably a muscle pattern your body does almost automatically. When you get a new toothbrush, you don't have to relearn how to brush your teeth; your body generalizes to using the new toothbrush without any difficulty.

When children look for, create, and analyze patterns and relationships, they can apply their reasoning processes in progressively abstract ways. Algebraic thinking begins in preschool when children practice recognizing and creating patterns in colors, shapes, sounds, and movements. Later in elementary school, children explore numerical relationships like those found in skip-counting. Older children explore functional relationships such as the number of bicycles in a bike rack and the corresponding number of wheels. They also use algebra to find missing addends (2 + ? = 7) and explore patterns in place value (*What must we add to 34 to get 44? To get from 44 to 144? To get from 144 to 1,144?*). As you can see, we cannot and need not wait until high school to engage children in algebraic thinking.

TEACHING TIP

Investigations Using Algebraic Thinking

While patterning has been de-emphasized in some early childhood curricula, algebraic reasoning remains a focus, particularly patterns in our number system. We want children to begin to generalize based on patterns they observe. One

(continued)

(continued)

favorite pattern in elementary school is odd and even numbers. Using a hundreds pocket chart with numbers in two colors (e.g., red and black) can help children in their investigations of the following conjectures:

- "If you add 1 to an even number, do you always get an odd number? Why do you think that is?"
- "If you add 2 to an odd number, do you always get another odd number? Why does that make sense?"
- "If you start at 1 and keep adding 2, do you get all the odd numbers? What pattern do you notice?"

Pairing the hundreds pocket chart with hands-on materials helps children begin to generalize the following:

- "If you can separate a number of objects into two equal groups with no leftovers, it's an even number. If one's left over, it's an odd number."

The notion that elementary school students can and should engage in algebraic thinking is being increasingly accepted and advocated for. When we offer materials that engage the algebraic thinking practices of generalizing, representing, justifying, and reasoning with mathematical relationships, we provide rich math-positive experiences that engage elementary school students in early algebra.

2. What does algebra look like in elementary school?

Algebra is a separate strand in mathematics; however, some standards embed the objectives for algebraic thinking within number and operations. In this way, algebraic thinking is applied to making sense of other areas of mathematics and real-world situations (Van de Walle, Karp, and Bay-Williams 2019). For a summary of the big ideas for algebra, see the following table. The summary illustrates how algebraic thinking combines the study of our number system, patterns and functions, and the use of symbols, variables, and expressions to model mathematical ideas (Stephens et al. 2015).

"PAUSE"ATIVE BOX

Summary of Big Ideas for Algebraic Thinking for Kindergarten through Grade 5

Here are some of the big ideas children will encounter when learning about algebraic thinking in school, including activities you can do to support their learning. These ideas are connected to the Common Core State Standards (National Governors Association Center for Best Practices and Council of Chief State School Officers 2010). Pause and take a moment to review the list. Put an asterisk next to some of the activities you'd like to try. Try them with your child. How did it go?

Grade	Big Ideas In Algebraic Thinking	Math Activities
K	Understand addition as adding to and subtraction as taking apart and taking from. • Using objects, fingers, mental images, drawings, sounds, acting out situations, verbal explanations, expressions, or equations to represent problems (within 10). • Use objects or drawings to find all possible sums for the number 10. • Fluently add and subtract within 5.	Make a *Ten Bead Bracelet* with children: Thread a pipe cleaner through ten same-colored beads and twist the ends to make a bracelet. See the sums of 10 by sliding the beads apart. The sums for 10 are: $0 + 10$ $1 + 9$ $2 + 8$ $3 + 7$ $4 + 6$ $5 + 5$ For practice with addition and subtraction, have children put five, two-color counters in a cup, shake, then spill them onto the table. Children tell an addition or a subtraction sentence to match the counters they see. For example, 3 red and 2 yellow counters could be thought of as:

(continued)

(continued)

Grade	Big Ideas In Algebraic Thinking	Math Activities
K		$3 + 2 = 5$, $2 + 3 = 5$, $5 - 3 = 2$, or $5 - 2 = 3$. For an added challenge, cover up one color of counters and have children guess how many are hidden.
	Decompose numbers using objects or drawings (within 10).	Explore decomposition with *Penny Baggies*. Draw a thick line down the center of a resealable baggie. Place five to ten pennies in the baggie. Shake, then lay the baggie on a table to see the addends. Talk with children about how a whole number can be broken into two parts (decomposed). Shake again. If desired, make an organized list of the wholes and parts you all discover. Your list might look like this: 6 → 3, 3 For more about number bonds, see Chapter 4, Question No. 6.
1	Addition and subtraction • Solve word problems using objects, drawings, and equations with a symbol for the unknown (within 20 and with up to three addends). • Understand and apply commutative and associative properties and the relationship between addition and subtraction to solve problems.	Addition and subtraction are best experienced through discussion, hands-on activities, and strong connections between the operations. A math storytelling strategy called *Looks and Talks* gets children telling math stories using part-part-whole thinking. Show children a picture that lends itself to part-part-whole discussion then ask, what math stories can you tell about this picture? See the picture on next page as an example.

(continued)

(continued)

Grade	Big Ideas In Algebraic Thinking	Math Activities
1 (cont'd)	• Understand subtraction as an unknown-addend problem. • Relate counting to addition and subtraction. • Fluently add and subtract within 10 using mental strategies. • Understand the equals sign means "is the same as" (e.g., $9 = 9$, $9 = 8 + 1$, $7 + 2 = 2 + 7$). • Find the unknown in an equation (e.g., $8 + \_\_ = 12$, $5 = \_\_ - 4$, $6 + 7 = \_\_$).	*There are 3 flowers. 2 are daisies and 1 is a tulip. There are 2 two-wheeled bicycles and 1 three-wheeled bicycle. There are 6 sections of sidewalk—2 with pebbles and 4 plain.* For an added challenge, make up an addition or subtraction problem to match the picture. Use hands-on objects to connect counting to subtraction by playing *Shrinking Train*. Have children make a train by connecting twenty linking cubes. Take turns rolling a die and removing that number of cubes from the train. The person who removes the last cube is the winner.
2	Addition and subtraction • Solve one- and two-step word problems using drawings and equations with a symbol for the unknown (within 100). • Fluently add and subtract within 20 using mental strategies. By end of grade 2, know from memory all sums of two one-digit numbers.	*The Sum What Dice Game* (Stenmark, Thompson, and Cossey 1986) gives practice with addition facts and mental math. Have children write the digits 1 through 9 on a paper. | 1 | 2 | 3 | 4 | 5 | 6 | 7 | 8 | 9 | Take turns rolling two dice. On each turn, the player may cover either the sum rolled on the dice or any two numbers that are still uncovered and that add to the sum rolled. For example, if a sum of 8 is rolled first, the player may cover: 8, or 1 and 7, or 2 and 6, or 3 and 5. Later in the game, if the sum of 8 is rolled again and the 5 is already covered, then the player cannot use the 3 and 5 combination and must play one of the other open possibilities.

(continued)

(continued)

<table>
<tr><th>Grade</th><th>Big Ideas In Algebraic Thinking</th><th>Math Activities</th></tr>
<tr><td rowspan="2">2</td><td></td><td>When a player cannot play, they are out and have a score of the sum of the uncovered numbers. Play continues for remaining players until everyone is out. The last person to go out will not necessarily win; the person with the lowest score wins.</td></tr>
<tr><td>Multiplication
• Determine if a set (up to 20) has an odd or even number of objects by pairing.
• Use repeated addition to find the total number of objects arranged in rectangular arrays (up to 5 by 5).</td><td>Odd and Even Bean Game helps children see patterns in our number system. Have children take a small handful of up to twenty dry beans. Work together to pair the beans up in groups of two. Count by twos to find the total number of beans. If there is a bean left without a partner, the number of beans is odd. If there are no leftovers, the number is even. Make a table and color the odd numbers yellow and the even numbers red. Talk about the patterns you see.
<table><tr><td>1</td><td>2</td><td>3</td><td>4</td><td>5</td><td>6</td><td>7</td><td>8</td><td>9</td><td>10</td></tr><tr><td>11</td><td>12</td><td>13</td><td>14</td><td>15</td><td>16</td><td>17</td><td>18</td><td>19</td><td>20</td></tr></table></td></tr>
<tr><td>3</td><td>Multiplication and division
• Solve equal-sized groups multiplication problems (e.g., 5 × 7 as 5 groups of 7 objects each) and arrays (within 100).
• Solve partitive and measurement division problems.
• Find the unknown in an equation relating three numbers (e.g., 8 × ? = 56, 15 = __ ÷ 3, 6 × 7 = ?).
• Apply properties of operations (commutative, associative, and distributive) as strategies to multiply and divide.</td><td>Several picture books offer engaging introductions to multiplication and division. Here are a few of my favorites to share with children:
• Bean Thirteen (McElligott 2007)
• The Lion's Share (McElligott 2012)
• 365 Penguins (Fromental 2017)
• Math Potatoes (Tang 2005)
Develop understanding of the commutative property of multiplication by playing Commutative Bingo. You will need a Bingo board like the one shown on the following page.</td></tr>
</table>

(continued)

(continued)

<table>
<tr><th>Grade</th><th>Big Ideas In Algebraic Thinking</th><th>Math Activities</th></tr>
<tr><td rowspan="2">3</td><td>
<ul>
<li>Understand division as an unknown-factor problem.</li>
<li>Fluently multiply and divide within 100 using mental math strategies. By the end of third grade, know from memory all products of two one-digit numbers.</li>
</ul>
</td><td>
<table>
<tr><td>×</td><td>4</td><td>5</td><td>6</td><td>7</td><td>8</td><td>9</td></tr>
<tr><td>4</td><td></td><td></td><td></td><td></td><td></td><td></td></tr>
<tr><td>5</td><td></td><td></td><td></td><td></td><td></td><td></td></tr>
<tr><td>6</td><td></td><td></td><td></td><td></td><td></td><td></td></tr>
<tr><td>7</td><td></td><td></td><td></td><td></td><td></td><td></td></tr>
<tr><td>8</td><td></td><td></td><td></td><td></td><td></td><td></td></tr>
<tr><td>9</td><td></td><td></td><td></td><td></td><td></td><td></td></tr>
</table>
Make two sets of cards numbered 4, 5, 6, 7, 8, 9 for a total of 12 cards.
Each player uses their own Bingo board. Pick two cards and multiply the numbers. If you pick a 6 and a 5, you can decide whether to write the product, 30, in the 5 × 6 box or the 6 × 5 box, but not both. Put the cards back in the pile. The object of the game is to get 6 in a row horizontally, vertically, or diagonally.
To find out more about the properties of multiplication and division, see Question No. 7 in this chapter. To learn about partitive and measurement division, see Chapter 4, Question No. 17.
</td></tr>
<tr><td>
Algebraic thinking
<ul>
<li>Solve two-step word problems using +, =, ×, ÷. Represent these problems using equations with a letter standing for the unknown.</li>
<li>Evaluate the reasonableness of answers using mental computation, estimation, and rounding.</li>
<li>Identify arithmetic patterns such as those in addition or multiplication grids and explain them using properties of operations.</li>
</ul>
</td><td>
Try this activity to find patterns in the multiplication chart. Draw an 11 × 11 grid and label it as shown here.
<table>
<tr><td>×</td><td>1</td><td>2</td><td>3</td><td>4</td><td>5</td><td>6</td><td>7</td><td>8</td><td>9</td><td>10</td></tr>
<tr><td>1</td><td></td><td></td><td></td><td></td><td></td><td></td><td></td><td></td><td></td><td></td></tr>
<tr><td>2</td><td></td><td></td><td></td><td></td><td></td><td></td><td></td><td></td><td></td><td></td></tr>
<tr><td>3</td><td></td><td></td><td></td><td></td><td></td><td></td><td></td><td></td><td></td><td></td></tr>
<tr><td>4</td><td></td><td></td><td></td><td></td><td></td><td></td><td></td><td></td><td></td><td></td></tr>
<tr><td>5</td><td></td><td></td><td></td><td></td><td></td><td></td><td></td><td></td><td></td><td></td></tr>
<tr><td>6</td><td></td><td></td><td></td><td></td><td></td><td></td><td></td><td></td><td></td><td></td></tr>
<tr><td>7</td><td></td><td></td><td></td><td></td><td></td><td></td><td></td><td></td><td></td><td></td></tr>
<tr><td>8</td><td></td><td></td><td></td><td></td><td></td><td></td><td></td><td></td><td></td><td></td></tr>
<tr><td>9</td><td></td><td></td><td></td><td></td><td></td><td></td><td></td><td></td><td></td><td></td></tr>
<tr><td>10</td><td></td><td></td><td></td><td></td><td></td><td></td><td></td><td></td><td></td><td></td></tr>
</table>
</td></tr>
</table>

(continued)

(continued)

Grade	Big Ideas In Algebraic Thinking	Math Activities
3		Begin by filling in the row and column for multiplying by 1. Ask children, what do you notice? This shows the identity property for multiplication. $n \times 1 = n$ Fill in the row and column for the twos. What can you say about these numbers? They are all even. Next fold the paper along the diagonal beginning from the multiplication symbol in the top left corner. Make a heavy crease. Fill in the boxes above the crease. Look at the boxes below the crease. How do they relate to the area above the crease? They are the same products in reverse order. This shows the commutative property of multiplication. What other patterns can you find in the multiplication chart? Can you find doubles and square numbers? Don't forget to look along the diagonals.
4	Use +, =, ×, ÷ to solve problems • Interpret a multiplication equation as a comparison (e.g., $63 = 9 \times 7$ means 63 is 9 times as many as 7 and 7 times as many as 9). • Multiply or divide to solve word problems involving multiplicative comparison (e.g., a 14-foot tall giraffe is two times as tall as an elephant. How tall is the elephant?). • Solve multistep word problems posed using +, =, ×, ÷, including problems with remainders. Represent problems using equations with a letter standing for the unknown.	In multiplicative comparison problems, we use the size of one group as a reference for a multiplicative idea. Here is an example: *Ty has 7 balloons. Julian has 3 times as many balloons as Ty. How many balloons does Julian have?* In this problem, the reference group is Ty's set of seven balloons. To learn more about multiplicative comparison problems, see Chapter 4, Question No. 15. Factors are all the positive numbers that can be multiplied to form a product. Another way of saying this is the factors of 24 are 1, 2, 3, 4, 6, 8, 12, and 24 because those are the numbers that evenly divide the number 24. The factor pairs are sets of two numbers that, when multiplied, will result in a given product. So the factor pairs for 24 are: 1×24 2×12 3×8 4×6

(continued)

(continued)

Grade	Big Ideas In Algebraic Thinking	Math Activities
4 (cont'd)	• Evaluate the reasonableness of answers using mental computation, estimation, and rounding. • Find all factors for numbers 1–100 and determine whether a number is prime or composite.	Play *Factor Game* on the National Council of Teachers of Mathematics website (search online for *Illuminations Factor Game*). This game provides practice relating multiplication and division and finding factors. Players take turns choosing numbers and coloring factors. Children can play against the computer or against another person.
	Algebraic thinking • Create number or shape patterns that follow a given rule. Identify apparent features of the pattern (e.g., odd or even, divisible by 5) and explain why features occur.	Here are a few interesting number patterns to talk about together: 1, 6, 11, 16, 21, 26 . . . 2, 7, 12, 17, 22, 27 . . . 3, 8, 13, 18, 23, 28 . . . What do you notice about the patterns individually? What do you notice about them collectively?
5	Use parentheses, brackets, or braces in numerical expressions, and evaluate expressions with these symbols.	Parentheses, brackets, and braces are all used to group operations. When simplifying an expression, the operations within these grouping structures are performed first. This is part of a larger idea called *order of operations*, an agreed-upon order for performing operations to simplify expressions. Try this experiment to see what happens when we do not complete operations within the groupings first: *Problem A* $(10 + 4) \times 2 =$ $14 \times 2 =$ Solution: 28 *Problem B* $10 + (4 \times 2) =$ $10 \times 8 =$ Solution: 80 Those grouping symbols make a big difference in the solution.

(continued)

(continued)

Grade	Big Ideas In Algebraic Thinking	Math Activities
5	Write simple expressions without evaluating them. For example, write *3 × (5 + 1)* to express "add 5 and 1, then multiply by 3." Recognize that 6 × (5,525 + 788) is six times as large as 5,525 + 788, without completing the calculation.	Practicing reading expressions aloud as if they were stories helps children understand what the grouping symbol does to the operation. For example, here is a story for the expression, *3 × (5 + 1)*: *There are 3 groups of colored balloons—5 red and 1 blue ballons in each group. This shows that the total number of balloons is 3 times the combined set of 6 balloons.*
	Algebraic thinking • Generate two number patterns using two given rules. Identify apparent relationships between corresponding terms and use those terms to form ordered pairs. Graph the ordered pairs on a coordinate plane.	A *coordinate grid* is used for locating points. An *ordered pair* names a point and gives its location on the coordinate grid. To plot a point, we need its location on the *x*-axis and the *y*-axis. Give this idea context and meaning by generating an input/output table with topics children care about. For more about input/output tables, see Question No. 3 in this chapter. Here are a few ideas to get you started: • wheels on a bicycle (1, 2), • wheels on a tricycle (1, 3), • seasons in a year (1, 4), and • fingers on a hand (1, 5). Use the ordered pairs from the table to plot the points on the coordinate grid. Graph paper comes in handy for this activity.

TEACHING TIP

Facilitating Algebraic Thinking Discussions

To engage students in algebraic reasoning, you need to select and present a cognitively demanding mathematical task that provides students with opportunities to look for patterns, generalize, and explain their reasoning. For a productive class discussion, all students must be able to enter the task and have ways to explain their thinking. Once you have laid the groundwork with a great task, try the following to facilitate algebraic reasoning during a whole-class discussion in which several students (or partners if the task was a group effort) share their thinking:

- Ask the class to identify common forms of algebraic reasoning (generalizing, finding a pattern, etc.) that they recognize in their classmates' work.
- Ask students to apply their own reasoning to the reasoning of others and say how their reasoning is similar or different.
- Ask if students agree or disagree (respectfully) with part of an explanation or a whole explanation and to provide mathematical reasoning for this.
- Provide space for students to ask classmates clarifying questions. Ask if someone can add on to a classmate's ideas, repeat what a classmate has said, or explain it in their own words.
- Have students classify which strategies go together and why.
- Provide space for students to comment on the algebraic reasoning strategies they found effective in justifying their responses to the task. They may generalize that some strategies were more efficient or accurate. They may see patterns or ways these strategies could be used.

Rich discussion results from careful planning by the teacher, a strong classroom community that supports collective student participation, and engagement with algebraic reasoning. It's not easy, but discussions *about* algebraic reasoning help to *build* algebraic reasoning. And that's worth all the effort!

3. Input/output tables are new to me. What do they mean?

Tables are a way of organizing information so that we can look for relationships and predict how patterns grow, decrease, or change in other ways. Since it seems to be never-ending, let's use the laundry I do each week to see patterns and make predictions. We can use a table to identify the rule: $n \times 1.5$ = loads of laundry. Don't let the equation scare you; n is the number of kids whose laundry needs to be done. Each kid needs about 1.5 loads done per week. So, all that the intimidating equation means is that you multiply the number of kids by 1.5 to see the total number of loads that need to be done. Easy peasy, right? Since laundry never ends, we can start to see generalizations. What if we had five cousins staying with us from out of town? How many loads would I need to do? What if all seventeen of the cousins from both sides of the family came for a visit? How many loads would that require? We don't have to keep adding columns to the table, because we can apply the rule $n \times 1.5$ to figure out how many loads of laundry would be generated by the kids.

Input = Number of Kids (n)	Output = Number of Loads of Laundry
1	1.5
2	3
3	4.5
4	6
5	7.5
6	9
7	10.5
8	12

An input/output table (function table)

After we make the table, we can graph the ordered pairs on a coordinate grid. If all of my children plus their seventeen visiting cousins insisted on clean clothes, my laundry escapades would look like this:

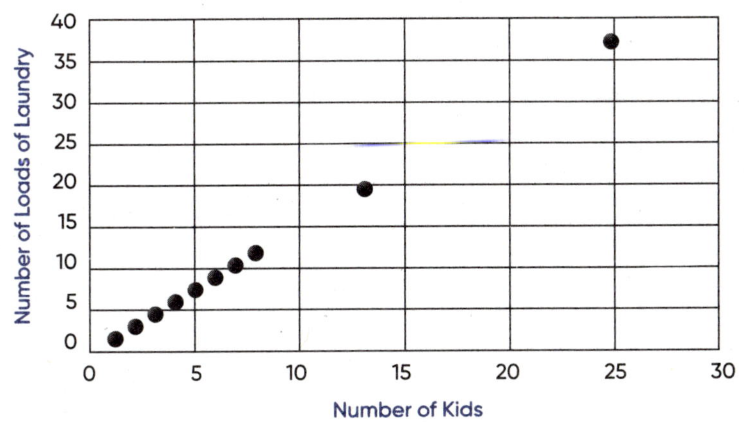

A coordinate grid with the ordered pairs from the input/output table graphed on it

4. What does it mean to *generalize*?

Generalizing means applying something known to a new situation or set of objects. Here is a conversation I had with my son Duncan when he was four years old in which he used algebraic reasoning to begin generalizing from examples:

Mom: Can you tell me about your pattern?

Duncan: It goes *car, car, truck, truck, car, car, truck, truck*.

Mom: I see. What vehicle will come next in the pattern?

Duncan: It will be a *car*.

Mom: How do you know?

Duncan: Because I already built it from the start, and I can see in my mind what will come next.

Mom: So, what if I changed the pattern to this: *bike, bike, tractor, tractor*. What would come next?

Duncan: It's a *bike*. It's just the same pattern only it has new vehicles.

Duncan generalized to a different set of toys and recognized how those toys could be used to make the same AABB pattern. This type of reasoning is fundamental as it helps children generalize arithmetic and notice patterns that hold true for algorithms and in the properties of the operations (Van de Walle, Karp, and Bay-Williams 2019).

An input/output table (function table) provides another visual for generalizing information. For example, we could use a table to find the number of wheels on Duncan's vehicles. If he has thirty-seven cars and trucks, the table helps us figure out that the numbers in the input column are multiplied by four to get the numbers in the output column. Knowing this helps us generalize and saves us from having to use repeated addition or fill out the whole table for all thirty-seven vehicles. We can "see" the pattern in the table.

Input = Number of Vehicles	Output = Number of Wheels
0	0
1	4
2	8
3	12
37	x

An input/output table (function table)

5. Why is the equals sign misunderstood and what can I do to help my child make sense of it?

In addition to generalization, algebraic reasoning requires concepts of equality, which we sometimes indicate with an equals (=) sign. However, children might view the equals sign as an operator sign, that is, a signal for "doing something" rather than as a relational symbol of equivalence or equal quantities.

Here are some of the misconceptions.

- Children might reject $5 = 5$ by saying there is no problem here.
- Children don't like $4 = 2 + 2$ because it just looks wrong—the answer is on the wrong side.
- When children see $4 + 3 = 5 + 2$ they may say that there are two problems but no answers.
- Children solve $5 + 14 = 19$ with no problem. But when they see $14 + ? = 19$, they often give 33 as the solution to the missing addend.
- Things get even trickier when an expression also includes a missing addend on the right side of the equals sign, like in this example: $5 + 14 = ? + 4$. Holy cow! Kids will often just grab all the numbers and add 'em up out of sheer frustration.

What's going on here? The issue stems from the meaning we assign to the equals sign. In the case of the problem $5 + 14 = ?$ the equals sign can be thought of as "the result of the previous computation." That is an accurate interpretation in *this* problem. However, in the example $4 + 3 = 5 + 2$, the equals sign must be interpreted differently. It is now a statement of equivalence between two quantities. And for the problem $5 + 14 = ? + 4$, we must find a number that is equal to 4 less than the sum of $5 + 14$. For goodness sake! The understanding that the equals sign requires that one side of the expression be equivalent to the other is a basic tenet of algebra.

"PAUSE"ATIVE BOX

Thinking about Math Language

Take a moment to pause and consider how you could support children's use of precise mathematical language. Your children's teachers may be stretching their algebraic reasoning skills by giving a variety of problems with unknowns in different positions, such as:

$$4 + ? = 17 + 2$$

$$? + 15 = 12 + 32$$

$$13 + 24 = 30 + ?$$

You can help out with this at home by using the phrase "has the same value as" instead of "makes" as you read number sentences aloud with your child. Try this activity: Write a few addition number sentences on a sheet of paper. If you write, for example, $2 + 2 = 4$, read the equation as "two plus two equals or has the same value as four." The term *makes* encourages the misconception that the equals sign is an action or an operation rather than representative of a relationship, whereas "is the same as" signals that the equation must be balanced on either side of the equals sign. You might read the missing addend problems shown in this box, as follows:

- What number, added to 15, would make each side of the equation have the same value as 12 plus 32?
- What number, added to 30, would make each side of the equation have the same value as 13 + 24?

Using the phrase "has the same value as" reinforces the concept of the equals sign as a balance with both sides of the equation requiring the same value.

Is It Balanced? An Activity for Understanding the Equals Sign

TEACHING TIP

A balance scale can help students develop the correct meaning of the equals sign. Try this activity with your class; you'll simply need a sticky note, pan balance, and counters.

1. Show students an equals sign you have drawn on a sticky note. Stick the equals sign in the center of the base of a pan balance and explain to students that the equals sign means that the items on one side of the pan balance must be equal to the items on the other side of the balance. That makes the balance, well, balanced.
2. Place three counters on one side of the scale. Place seven counters on the other side. Ask students, "Is it balanced? How many counters need to be removed (or added) to make them equal?"
3. Continue exploring and discussing the results of adding and removing counters to balance the scale.
4. You may choose to use cards with missing addend number sentences for students to model using the counters and pan balance. Continually reinforce by asking questions like:
   a. "Are they equal?"
   b. "Are they balanced?"
   c. "Do both sides of the equals sign have the same amount?"
   d. "How could we make both sides of the equals sign the same?"
5. When students have had ample experience with this exploration, have them draw pictures in their math journals to record their thinking about equality and the equals sign.

6. What other algebraic reasoning, besides generalization and equality, can I expect my child to encounter in elementary school?

Thinking about unknown quantities is the final element of algebraic reasoning children will be expected to understand as a foundation to the x, y, z of formal algebra. Along with the word *variable*, *unknown* is one of the words most frequently associated with algebra. You do algebra all the time without realizing it. Any time you figure out what time you need to leave to get to work on time, you are solving for an unknown and doing algebra. For example, if you have to be at work at 9:00 a.m. and it takes 45 minutes to get there, what time do you have to leave? This is an algebraic idea that could be represented like this:

$$? + 45 \text{ minutes} = 9 \text{ o'clock}$$

You could use a number line to solve it by jumping back 45 minutes like this:

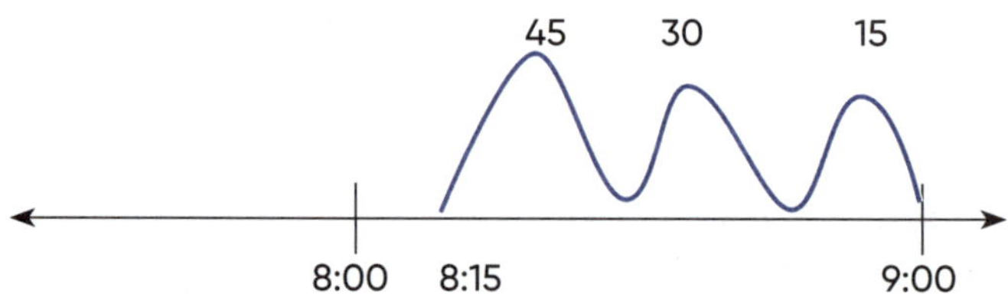

When solving for an unknown, algebraic reasoning comes into play as children use strategies like working backward, drawing a model, or using a complementary operation (subtraction in this case) to figure out that you have to leave the house at 8:15 a.m. in order to get to work on time. Word problems are great for helping children understand what's really happening in a missing addend number sentence like this:

$$? + 5 = 12$$

Some new baby ducks hatched at the farm. There were five ducks on the farm, and now there are twelve ducks. How many new baby ducks were hatched?

Baby ducks are interesting, but sometimes you gotta get real with algebra, right? Here's a context that will ring true to most children:

$$\$0.88 + ? = \$1.44$$

Tillie had \$0.88. She cleaned out the minivan and found some loose change in the seats and cup holders. Now she has \$1.44. How much money did Tillie find in the minivan?

We might choose to use a number line to solve this one. Big jumps are for showing counting by tens. We do this first because it's efficient. The little jumps show counting by one. It takes five big jumps (50) and six little jumps (6): 50 + 6 = 56.

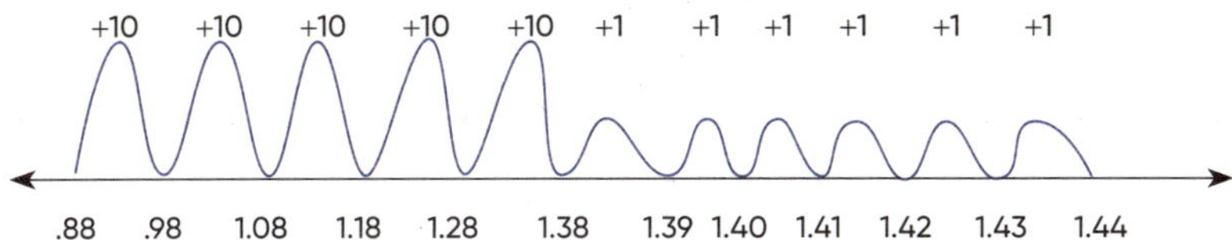

Solution: She found \$0.56 in the minivan.

Or this one:

$$75 + 39 = ? + 100$$

Nico had 75 cents in quarters and 39 cents in loose change. He made a dollar by combining his three quarters with 25 cents from his loose change. How much loose change does he have now?

A bar model helps us see how Nico could solve for the unknown.

75	39
100	?

In a bar model, each bar represents the values on either side of the equals sign. We can use this visual representation to demonstrate that 100 and the unknown value together make 75 + 39.

$$75 + 39 = 114$$
$$114 - 100 = 14$$

Solution: We recognize the value of 14 would make a true statement if substituted for the unknown.

When we solve for an unknown, we are using algebraic reasoning to work flexibly with numbers and to see ways we use algebra to solve everyday problems. These are foundational ideas that prepare students for formal algebra where they will analyze change and reason more abstractly and quantitatively.

7. What's the importance of the properties of the operations?

Understanding and using the properties of the operations eliminates lengthy computations, greatly reduces the number of facts that need to be drilled over and over, and gives children a reasoning strategy that can be generalized across many examples.

"PAUSE"ATIVE BOX

The Properties of the Operations

For a refresher on the properties of the operations, see the table here. Take a moment to pause and consider how each property helps us solve problems efficiently and flexibly. It's not enough to identify the property by its name or symbolic representation, though these are also given here. We must work to help children recognize how applying these properties makes math make sense. Try the suggested activities with children and watch their math-positive mindsets about the structure of our number system grow.

Property	Symbols We Use	Kid-Friendly Language or Example	Math Activities
Addition			
Identity	$a + 0 = a$	When you add zero to a number, the number stays the same.	Play the *Plus Zero* Game, which shows how adding 0 does not change the number in a set. For this game, you will need a traditional die and a 0 die. Make a 0 die by covering a die with masking tape and writing 0 on each face. Roll both dice and create sets for the two numbers shown on the dice. Children will soon catch on that the 0 die never needs a set of counters and doesn't change the number of counters in the other set.
Commutative	$a + b = b + a$	Changing the order of the addends doesn't change the sum.	Use a pan balance (instructions for a simple homemade balance can be found online) to explore commutativity. Have children place seven counters—three red and four blue—on one side of the balance. Have them place four red and three blue counters on the other side. Talk about how the two sides are balanced even though the colors are reversed.

(continued)

(continued)

Property	Symbols We Use	Kid-Friendly Language or Example	Math Activities
Associative	$(a + b) + c = a + (b + c)$	When you add three numbers, you can group the numbers any way and the sum is the same.	Use a pan balance and linking cubes to explore how grouping numbers doesn't change the sum. Make a stick representing (7 + 3) + 2 by joining seven red to three blue and placing them in the balance. Then place two individual yellow cubes into the balance. On the other side of the balance, place seven individual red cubes and a stick of three blue and two yellow to represent 7 + (3 + 2). Talk about how the two sides of the balance are equal.
Additive Inverse	$a - a = 0$ $a + (-a) = 0$	When you subtract a number from itself, you get zero.	Play *Go Back to Zero* to learn about the effect of subtracting a number from itself. Roll a die and build a set of counters for the number shown. Then call out, "Go back to zero!" Children must show what they do with their counters to get back to 0. They can count back or simply subtract the number shown on the die.
Inverse Relationship of Addition and Subtraction	*If* $a + b = c$ *then* $c - b = a$ *and* $c - a = b$.	23 – ? = 17. Think, 17 plus what number equals 23?	Write this set of problems on a sheet of paper: 17 + 23 = 40 – 23 = 40 – 17 = 23 + 17 = Have children tell what they notice about the problems. They should see that all the problems contain the same numbers but different operations. Ask if the problems can be solved without doing any calculating. Talk about how the problems are related. If children look closely, they will notice the solutions can be found in the related problems due to the inverse relationship between addition and subtraction. Use counters in two different colors to reinforce this idea. Show how sticks showing the four equations are all the same length, 40 counters long. The parts are the same too, 23 and 17.

(continued)

(continued)

Property	Symbols We Use	Kid-Friendly Language or Example	Math Activities
Multiplication			
Zero	$a \times 0 = 0$	When you multiply a number by 0, the product is 0.	Children need a lot of discussion and hands-on experience to understand this property, because it is often confused with the identity property of addition. Work with children to build sets of counters that show what's going on in silly word problems like these: • *I have 10 packages of gum with 0 sticks in each package. How much gum do I have?* • *I bought 8 cases of candy bars with 0 candy bars in each case. How many candy bars do I have?* • *I collected 0 packs of baseball cards that come in packs of 5. How many baseball cards do I have?* These are silly because they show how having zero packages and having lots of empty packages both give you a big fat zero.
Identity	$a \times 1 = a$	When you multiply a number by 1, the number stays the same.	Explore how multiplying by one does not change the number by playing the *Mystery Number* game. Tell children you are hiding a number behind your back. Tell them to roll a die and you will give the product of the number on the die multiplied by the mystery number. Be as quick as you can in supplying the product. After a lot of rolls and giggles, ask children, "What is the mystery number?" Children will soon catch on that products of 1 and any number can be figured in an instant.

(continued)

(continued)

Property	Symbols We Use	Kid-Friendly Language or Example	Math Activities
Commutative	$a \times b = b \times a$	When you multiply two numbers in any order, the product is the same.	Arrays and area models are great for exploring the commutative property. Play *How Long, How Many* using Cuisenaire rods (printable rods can be found online) and a 10-by-10 grid of centimeter paper (also available online). Roll a die. This is *How Long* of a rod you should take. Roll the die again. This is *How Many* rods you should take. Arrange the rods on the grid paper in a rectangle and trace around them. Remove the rods and write a multiplication problem to label the rectangle. For example, if you roll a 6 and a 3, you would find the rod six units long and use three of them to make a rectangle. Label it 3×6, meaning 3 groups of rods that are 6 units long. The object of the game is to be the player who has covered the most squares after five turns. If you cannot place a rectangle because you do not have enough space, you lose a turn. What strategies did you find for placing the rectangles? What does this show about the commutative property? For more practice with the commutative property, play *Commutative Bingo* shown in the Summary of Big Ideas for Algebraic Thinking in this chapter.
Associative	$(a \times b) \times c = a \times (b \times c)$	Changing the grouping of the numbers doesn't change the product.	Area models work well for showing the associative property. Draw rectangles on grid paper to compare how two groupings of multiplication problems have the same combined area. One person represents $(2 \times 3) \times 4$ by drawing a 2×3 rectangle four times and counting the total number of grid squares in the rectangles. Now represent $2 \times (3 \times 4)$ by drawing a 3×4 rectangle two times on grid paper. Count the total number of squares. How do they compare? Will this work every time? Try it and talk about why the rectangles look different but have the same total area.

(continued)

(continued)

Property	Symbols We Use	Kid-Friendly Language or Example	Math Activities
Inverse Relationship of Multiplication and Division	*If* $a \times b = c$ *then* $c \div b = a$ *and* $c \div a = b$.	$48 \div ? = 12$. Think, 12 multiplied by what number equals 48?	Number bonds for multiplication and division help children see how these operations are the inverse (opposite) of one another. Instead of learning all the multiplication facts then moving on to division, experts recommend starting division pretty soon after learning multiplication facts, giving children a lot of experience with the strong relationship between these two operations (Van de Walle, Karp, and Bay-Williams 2019).
Distributive	$a \times (b + c) = a \times b + a \times c$	$6 \times 24 = ?$ is not a fact I know. So, I can break the 24 into parts that I can solve: $6 \times (20 + 4) = ?$ Now I add them together. $120 + 24 = 144$.	A multiplication problem can be represented using a rectangle. That rectangle can be partitioned (divided) and then represented as two rectangles while keeping the same value. This is the distributive property and we use partial products (easier facts) to find the answer. For example, in the image here, we see how the problem 6×24 can be partitioned to make it easier to solve. We multiply 6×20 then 6×4 and add the partial products to find the answer. 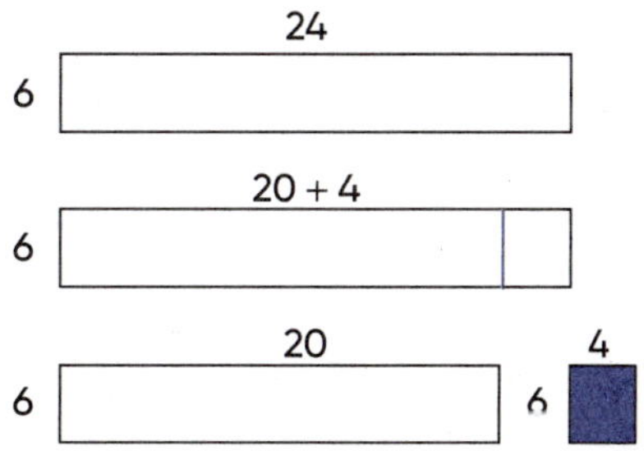

8. Why is it important to learn about both repeating and growing patterns?

Patterns are found all around us, from weather to economics to music to trends in hair styles. Recognizing, creating, extending, completing, and translating patterns are all part of algebraic thinking. Repeating patterns are those with a core that is duplicated. When your child sets the table for dinner, the repeating pattern is *fork, plate, knife, spoon; fork, plate, knife, spoon*. Growing patterns involve a progression or growing sequence. Think, "There Was an Old Lady Who Swallowed a Fly." The old lady gulps a fly, then a spider to catch the fly, then a bird to catch the spider to catch the fly, then a cat to catch the bird to catch the spider to catch the fly. Just like there are growing patterns, there are decreasing patterns. The song, "B-I-N-G-O" exemplifies a decreasing pattern, with one letter dropping off each time the verse is repeated. We often record patterns symbolically using letters, which allows us to translate patterns to new materials or contexts. If you used symbols to represent the growing pattern in "There Was an Old Lady Who Swallowed a Fly," it could look like this: A, AB, ABC, ABCD, and continuing.

"PAUSE"ATIVE BOX

Repeating and Growing Patterns

Take a moment to pause and think about the figures presented here. A repeating AAB pattern is represented using movements (clap, clap, snap) and shapes (rectangle, rectangle, triangle). Now look at the growing pattern. It increases by two rhombi with each successive step. Where do you think the two new rhombi will be placed in Step 4? Beginning in preschool, children explore patterns that repeat and patterns that grow. What are some repeating patterns in your math-focused home or classroom? The daily classroom schedule, a weekly soccer practice or piano lesson, and monthly trips to the orthodontist are repeating patterns. With a repeating pattern, we can predict what comes next based on a core. The core is the smallest, repeated unit in the pattern. For the repeating pattern example here, the core is clap, clap, snap. On the other hand, a growing pattern does just that—it grows. The focus is on analyzing how the pattern is changing with each subsequent step. Growing patterns are functions—the number of miles added to your car's odometer each week is a function of how far you drive. Whether it's counting months until a birthday or watching a college savings account grow, patterns help us anticipate and generalize to understand mathematics better.

Repeating Patterns

Clap, clap, snap, clap, clap, snap

A A B A A B

Growing Patterns

Step 1 Step 2 Step 3

9. I still get algebra angst. What are some materials I can use to help my child understand algebra concepts?

Though algebra concepts begin quite simply, generalizations and symbols quickly overtake the colorful patterns created by a preschooler. Young children should explore patterns, functions, and number relationships with hands-on materials to solidify their age-appropriate algebraic thinking. To help a child build algebraic thinking, a classroom (or a math-focused home) might be stocked with a variety of items like those that follow.

"PAUSE"ATIVE BOX

Manipulatives for Use with Algebraic Thinking

Following is a list of manipulatives most frequently stocked in classrooms for supporting the learning of algebraic thinking. Take a moment to pause and review the list. I've also given a suggestion for a simple task using each type of manipulative. Put an asterisk by the activities you'd like to try with your child. Try them out. How did it go? Have fun building a math-positive mindset about algebra along with your children.

Manipulative	Description	Uses in Algebraic Thinking	Math Activities
Pattern Blocks	Plastic or wooden blocks that come in six shapes: yellow hexagons, orange squares, red trapezoids, green triangles, tan rhombi, and blue rhombi.	Sorting, classifying, and working with patterns in shape and color.	Use pattern blocks to create repeating patterns. Identify the core of the pattern. Have children cover their eyes while you remove a step from the center of the pattern. Have children fill in the missing step.
Color Cubes	Sometimes called *inch cubes*, these foam, plastic, or wooden cubes are handy for building algebraic reasoning.	Exploring patterns and equations	Use color cubes to create repeating and growing patterns. For growing patterns, build the first three steps of the pattern, then have children use reasoning and visualization to describe what the pattern would look like if extended to Step 10. How many blocks would there be? What would the design look like? Then build it together using blocks, a drawing, or an equation.

(continued)

(continued)

Manipulative	Description	Uses in Algebraic Thinking	Math Activities
Dice	Random number generators. Can be programmed to use any numbers, shapes, or operations by covering the sides with masking tape and writing on the faces.	Generating numbers for equations	Have children use four dice to create a number sentence or an inequality using an operation of their choice. They can turn the dice to any side they wish. An example might be: $6 + 4$ is less than $6 + 6$ Then have them try to trick you by writing three true equations and one false equation. Which of these equations is false? 1×6 equals 6×1 2×3 equals 6×1 3×3 is greater than 1×6 3×4 is not equal to 2×6
Pan Balance	Sometimes called a *balance scale* or a *bucket balance*, a pan balance has a beam with two cups or pans on either end. When the cups contain the same weight, the beam is balanced.	Exploring equations	For young children, place seven cubes on one side of the balance and five cubes on the other. Ask, "How many cubes need to be added (or taken away) to make the scale balanced?" For older children, place a banana *(b)* on one side of the balance and have children determine how many grapes *(g)* it takes to make the scale balanced. Use symbols to record the equation: $b = 23g$. This equation means 1 banana equals 23 grapes (if we're comparing their weight).

(continued)

(continued)

Manipulative	Description	Uses in Algebraic Thinking	Math Activities
Pan Balance (*continued*)			Use other household items and record the equations using symbols. Remember to emphasize that the equals sign means "is the same as." Try the Pan Balance games on the National Council of Teachers of Mathematics website (search online for *Illuminations Pan Balance*). These games provide a great visual of the meanings of equivalence, value, and balance.
Hundreds Chart	Grid listing numbers from 0 to 99 or 1 to 100.	Exploring patterns in our number system	Print out several copies of a completed hundreds chart (available online). On separate charts, have children color in the numbers for skip counting by twos, threes, and fives and compare the patterns they see. What do they notice and wonder about these patterns? With a new chart, have children identify a mystery number by coloring in all the numbers it could *not* be. Give clues like this: • The number is even. • The number is greater than 50. • The digits in the number sum to 12. • The number is less than 80. Answer: 66

QUICK REFERENCE CHART

Common Concerns about Algebra

This table is a quick reference for some of the *most common concerns* I hear from parents. The table lists the concern, an explanation of the thinking behind it, and an idea that addresses the concern. Consider photocopying this table and placing it somewhere where you can be reminded of what is needed to build your math-positive mindset and that of your children and/or students.

Concern	Explanation	What to Do	Question Number(s) (to Learn More)
Why algebra?	Algebraic reasoning is all about identifying relationships and generalizing those ideas to new situations or objects. Our base ten system is one of the first algebraic ideas children encounter. It's a growing pattern that never ends!	Display a hundreds chart in children's bedrooms to prompt the discovery of patterns in our number system. What do they see as they look down the columns or along a diagonal? What other interesting patterns can they discover? How are these patterns made?	1
What are algebra concepts for elementary school?	Preschoolers recognize and create patterns in colors, shapes, sounds, and movements. Elementary-age children explore numerical relationships like those found in skip-counting. Older children explore functional relationships such as the number of legs on a farm with 12 chickens and 8 lambs.	Use your own family as fodder for algebraic discussions. How many eyes in our family? We just multiply the number of people in our family by two, because each person has two eyes. How about toes? Is there a pattern? Algebra combines the study of our number system, patterns and functions, and the use of symbols, variables, and expressions to model mathematical ideas.	2

(continued)

(continued)

Concern	Explanation	What to Do	Question Number(s) (to Learn More)
How can I support algebraic understanding in a math-focused home?	Appreciating the importance of algebra comes from recognizing its usefulness in many areas of mathematics and the real world.	Algebra connects to all other areas of math, but children probably won't see these connections without your help. When exploring the properties of operations, connect children's work with basic facts. When creating repeating patterns with pattern blocks, remind children how they use these tools in geometry too to create designs that have symmetry (another repeating pattern). When filling in an input/output table, talk about how algebra helps with skip-counting. Connections are essential to helping children see the threads of mathematics that extend across content areas and processes.	2
Input/output tables	Input/output tables show how two pieces of information are dependent on one another. For example, one bicycle has two wheels, two bicycles have four wheels, and so on. The number of wheels is dependent on the number of bicycles.	Just like in the previous example, it's fun to use your own family for creating input/output tables. The number of eyes and toes can be displayed in a table. You can then have children generalize to see how many eyes or toes there would be if your family grew to 99 members.	3

(continued)

(continued)

Concern	Explanation	What to Do	Question Number(s) (to Learn More)
What does *generalize* mean?	Generalizing means applying something known to a new situation or set of objects. This type of reasoning is fundamental as it helps children generalize arithmetic and notice patterns that hold true for algorithms and in the properties of the operations (Van de Walle, Karp, and Bay-Williams 2019).	Generalizing helps children use patterns to find answers to problems several steps away. *Bernardo is going on a bike trip. If he travels 1.5 miles every 20 minutes, how far will he travel in 2 hours?*	4
The equals sign	Algebraic reasoning requires concepts of equality, which we sometimes indicate with an equals (=) sign. Children might view the equals sign as an operator sign, that is, a signal for "doing something" rather than as a relational symbol of equivalence or equal quantities.	Use the phrase "has the same value as" instead of "makes" as you read number sentences aloud with your child. Read the equation $2 + 2 = 4$ as "two plus two equals or has the same value as four."	5

(continued)

(continued)

Concern	Explanation	What to Do	Question Number(s) (to Learn More)
Unknown quantities	Along with the word *variable*, *unknown* is one of the words most frequently associated with algebra. Children will be expected to understand such as a foundation to the *x*, *y*, *z* of formal algebra. When we solve for an unknown, we are using algebraic reasoning to work flexibly with numbers and to see ways we use algebra to solve everyday problems.	You do algebra all the time without realizing it. Any time you figure out what time you need to leave to get to work on time, you are solving for an unknown and doing algebra. Word problems are great for helping children understand what's really happening in a missing addend number sentence. For example: *Keyara has $75 dollars saved for a new bike. How much more does she need to save if the bike costs $135?* We can think of this problem as: $75 + x = 135$ (75 plus some amount has the same value as 135) or $135 - 75 = x$ (135 minus 75 has the same value as what number)	6

(continued)

(continued)

Concern	Explanation	What to Do	Question Number(s) (to Learn More)
Properties of operations	Understanding the properties of operations eliminates lengthy computations, greatly reduces the number of facts that need to be drilled, and gives children a reasoning strategy that can be generalized across many examples.	Do the activities listed in the table in Question No. 7, then celebrate by eliminating any flash cards that can be "known" by understanding a property. For example, when children can fully explain and show the meaning of the commutative property, celebrate by tossing out half the flashcards in the addition facts set and/or the multiplication facts set. Whew! Now we only need to work on half as many facts. How many other facts can be "understood" rather than memorized when we apply the properties of addition and multiplication? Try it out.	7
Repeating and growing patterns	A *repeating pattern* has a discernible repeated core. AAB, AAB, AAB is a repeating pattern. A *growing pattern* continues without a repeating core. A, AA, AAA, AAAA, AAAAA is an example.	Patterns are fun when they are tied to movement. Try this one: Have your child choose three movements. For example, *kick*, *stomp*, *shake*. Repeat the same sequence of the movements while saying the words. After a while say, "I'm replacing *kick*, *stomp*, *shake* with *A*, *B*, *C* because my mouth is getting tired." This connects the abstract symbols we commonly use to label patterns with the movements.	8

(continued)

(continued)

Concern	Explanation	What to Do	Question Number(s) (to Learn More)
Algebra tools to have on hand	Children should explore patterns, functions, and number relationships with hands-on materials to solidify their algebraic reasoning. Pattern blocks, color cubes, dice, pan balances, and hundreds charts are a few of my favorites for building algebraic reasoning.	Try using pattern blocks to build repeating and growing patterns. A pan balance comes in handy for exploring the true meaning of the equals sign. The hundreds chart shows many patterns in our counting system. Keeping experiences hands-on and meaningful ensures your child will have a "functional" relationship with algebra. (Aren't math puns fun?)	9

Bibliography

Adler, David A. 2015. *Triangles*. Illus. Edward Miller. New York: Holiday House.

Baroody, Arthur J., and Jesse L. M. Wilkins. 1999. "The Development of Informal Counting, Number, and Arithmetic Skills and Concepts." In *Mathematics in the Early Years*, edited by Juanita V. Copley, 48–65. Washington, DC: National Association for the Education of Young Children.

Beilock, Sian L., Elizabeth A. Gunderson, Gerardo Ramirez, and Susan C. Levine. 2010. "Female Teachers' Math Anxiety Affects Girls' Math Achievement." *Proceedings of the National Academy of Sciences* 107: 1860–63.

Boaler, Jo. 2014. "Research Suggests Timed Tests Cause Math Anxiety." *Teaching Children Mathematics* 20: 469–74.

———. 2013. "Ability and Mathematics: The Mindset Revolution That Is Reshaping Education." *FORUM* 55: 143–151.

———. 2012. "Timed Tests and the Development of Math Anxiety." *Education Week*, July 3. Accessed May 27, 2017. www.edweek.org/ew/articles/2012/07/03/36boaler.h31.html?print=1.

———. 1998. "Open and Closed Mathematics: Student Experiences and Understandings." *Journal for Research in Mathematics Education* 29: 41–62.

Burns, Marilyn. 2015. *About Teaching Mathematics: A K–8 Resource*. 4th ed. Sausalito, CA: Math Solutions.

———. 2004. "Writing in Math." *Educational Leadership* 62: 30–33.

———. 1984. *The Math Solution: Teaching Mathematics Through Problem Solving*. Sausalito, CA: Marilyn Burns Education Associates.

Burris, Justin T. 2013. "Virtual Place Value." *Teaching Children Mathematics* 20: 228–36.

Cai, Jinfa. 2010. "Helping Elementary School Students Become Successful Mathematical Problem Solvers." In *Teaching and Learning*

Mathematics: Translating Research to the Classroom, edited by Diana Lambdin, 241–54. Reston, VA: NCTM.

Cain, Judith S. 2002. "An Evaluation of the Connected Mathematics Project." *The Journal of Educational Research* 95: 224–33.

Carpenter, Thomas P., Elizabeth Fennema, Megan Loef Franke, Linda Levi, and Susan B. Empson. 2014. *Children's Mathematics: Cognitively Guided Instruction*. 2d ed. Portsmouth, NH: Heinemann.

Chapin, Suzanne H., Catherine O'Conner, and Nancy Canavan Anderson. 2013. *Talk Moves: A Teacher's Guide for Using Classroom Discussions in Math*, 3d ed. Sausalito, CA: Math Solutions.

Conklin, Melissa. 2010. *It Makes Sense! Using Ten-Frames to Build Number Sense*. Sausalito, CA: Math Solutions.

Constantino, Steve M. 2016. *Engage Every Family: Five Simple Principles*. Thousand Oaks, CA: Corwin.

Copley, Juanita. 2010. *The Young Child and Mathematics*. 2d ed. Washington, DC: National Association for the Education of Young Children; Reston, VA: National Council of Teachers of Mathematics.

Copple, Carol, and Sue Bredekamp. 2009. *Developmentally Appropriate Practice in Early Childhood Programs Serving Children from Birth Through Age 8*. 3d ed. Washington, DC: National Association for the Education of Young Children.

Cutler, Carrie S. 2007. *Calculator Activities: Measurement*. Illus. David Parker. Waco, TX: Prufrock Press.

———. 2005. "Knowing When to Which: Addressing the Calculator Quandary." *The Centroid: Official Journal of the North Carolina Council of Teachers of Mathematics* 31: 10–12.

Cutler, Carrie S., and Eula Ewing Monroe. 1999. "What Are You Learning, Billy Boy, Billy Boy? The Diary of a Teacher's Incorporation of Portfolios into Mathematics Instruction." *Contemporary Education* 70: 52–55.

Dawson, Peg. (n.d.). "Homework: A Guide for Parents." National Association of School Psychologists. Accessed May 27, 2017. http://www.naspcenter.org/home_school/homework.html.

Duncan, Greg, J. Chantelle J. Dowsett, Amy Claessens, Katherine Magnuson, Aletha C. Huston, Pamela Klebanov, Linda S. Pagani, Leon Feinstein, Mimi Engel, Jeanne Brooks-Gunn, Holly Sexton, Kathryn Duckworth, and Crista Japel. 2007. "School Readiness and Later Achievement," *Developmental Psychology* 43: 1428–46.

Dweck, Carol. 2006. *Growth Mindset: The New Psychology of Success.* New York: Ballantine.

Eisenhauer, Mary Jane, and David Feikes. 2009. "Dolls, Blocks and Puzzles: Playing with Mathematical Understandings." *Young Children* 64: 18–25.

Franke, Megan L., Elham Kazemi, and Angela Chan Turrou. 2018. *Choral Counting and Counting Collections: Transforming the PreK–5 Classroom*. Portsmouth, NH: Stenhouse.

Freer Weiss, Dana M. 2006. "Keeping It Real: The Rationale for Using Manipulatives in the Middle Grades." *Mathematics Teaching in the Middle School* 11: 238–42.

Fromental, Jean-Luc. 2017. *365 Penguins*. Illus. Joëlle Jolivet. New York: Harry N. Abrams.

Fuson, Karen, William M. Carroll & Jane V. Drueck. 2000. "Achievement Results for Second and Third Graders Using the Standards-Based Curriculum Everyday Mathematics." *Journal for Research in Mathematics Education* 31: 277–95.

Gerth, Melanie. 2000. *Ten Little Ladybugs*. Illus. Laura Huliska-Beith. Franklin, TN: Piggy Toes Press.

Ginsburg, Herbert P. 2016. "Finding the Math in Storybooks for Young Children." *Mind/Shift* (podcast). https://ww2.kqed.org/mindshift/2016/02/02/finding-the-math-in-storybooks-for-young-children/.

Gunderson, Elizabeth A., Sarah J. Gripshover, Carissa Romero, Carol S. Dweck, Susan Goldin-Meadow, and Susan C. Levine. 2013. "Parent Praise to 1- to 3-Year-Olds Predicts Children's Motivational Frameworks 5 Years Later." *Child Development* 84: 1526–41.

Hamilton, William. 1859. "Lecture XIV, Consciousness: Attention in General." In *Lectures on Metaphysics and Logic*. Vol. 1, edited by H. L. Mansel and J. Veitch. Boston: Gould and Lincoln.

Henderlong, Jennifer, and Mark R. Lepper. 2002. "The Effects of Praise on Children's Intrinsic Motivation: A Review and Synthesis." *Psychological Bulletin* 128: 774–95.

Hendrickson, Scott, Daniel Siebert, Stephanie Z. Smith, Heidi Kunzler and Sharon Christensen. 2004. "Addressing Parents' Concerns about Mathematics Reform." *Teaching Children Mathematics* 11: 18–23.

Henry, Valerie J., and Richard S. Brown. 2008. "First-Grade Basic Facts: An Investigation into Teaching and Learning of an

Accelerated, High-Demand Memorization Standard." *Journal for Research in Mathematics Education* 39: 153–83.

Hiebert, James, and Donald A. Grouws. 2007. "The Effects of Classroom Mathematics Teaching on Students' Learning." In *Second Handbook of Research on Mathematics Teaching and Learning*, edited by Frank Lester, 371–404. Charlotte, NC: Information Age.

Hiebert, James, Thomas P. Carpenter, Elizabeth Fennema, Karen C. Fuson, Diana Wearne, Hanlie Murray, Alwyn Olivier, and Piet Human. 1997. *Making Sense: Teaching and Learning Mathematics with Understanding*. Portsmouth, NH: Heinemann.

Hoban, Tana. 1996. *Shapes, Shapes, Shapes*. New York: Greenwillow Books.

Hoover-Dempsey, Kathleen V., and Howard M. Sandler. 1997. "Why Do Parents Become Involved in Their Child's Education?" *Review of Educational Research* 67: 3–42.

Hutchins, Pat. 1967. *Rosie's Walk*. New York: Macmillan.

Institute for the Future and Dell Technologies. 2017. *Emerging Technologies' Impact on Society and Work in 2030: The Next Era of Human/Machine Partnerships*. Retrieved from https://www.delltechnologies.com/content/dam/delltechnologies/assets/perspectives/2030/pdf/SR1940_IFTFforDellTechnologies_Human-Machine_070517_readerhigh-res.pdf.

Kilman, Marlene. 1999. "Beyond Helping with Homework: Parents and Children Doing Mathematics at Home." *Teaching Children Mathematics* 6: 140–46.

Kling, Gina, and Jennifer M. Bay-Williams. 2015. "Three Steps to Mastering Multiplication Facts." *Teaching Children Mathematics* 21: 548–59.

———. 2014. "Assessing Basic Fact Fluency." *Teaching Children Mathematics* 20: 488–97.

Kohn, Alfie. 2001. "Five Reasons to Stop Saying 'Good Job!'" *Young Children* 56: 24–30.

Krulik, Stephen, and Jesse A. Rudnick. 1988. *Problem Solving: A Handbook for Elementary School Teachers*. Boston: Allyn & Bacon.

Levine, Susan C., Linda Whealton Suriyakham, Meredith L. Rowe, Janellen Huttenlocher, and Elizabeth A. Gunderson. 2010. "What Counts in the Development of Young Children's Number Knowledge?" *Developmental Psychology* 46:1309–19.

McElligott, Matthew. 2012. *The Lion's Share*. New York: Bloomsbury USA Children.

———. 2007. *Bean Thirteen*. New York: G.P. Putnam's Sons Books for Young Readers.

Mirra, Amy, ed. 2004. *A Family's Guide: Fostering Your Child's Success in School Mathematics*. Reston, VA: NCTM.

Monroe, Eula Ewing. 2006. *Math Dictionary: The Easy, Simple, Fun Guide to Help Math Phobics Become Math Lovers*. Honesdale, PA: Boyds Mills Press.

Monroe, Eula Ewing, and Terrell Young, eds. 2018. *Deepening Students' Mathematical Understanding with Children's Literature*. Reston, VA: NCTM.

Moyer-Packenham, Patricia S:, Gwenanne Salkind, and Johanna J. Bolyard. 2008. "Virtual Manipulatives Used by K–8 Teachers for Mathematics Instruction: Considering Mathematical, Cognitive, and Pedagogical Fidelity." *Contemporary Issues in Technology and Teacher Education* 8: 202–18.

Moyer-Packenham, Patricia. S., and Arla Westenskow. 2013. "Effects of Virtual Manipulatives on Student Achievement and Mathematics Learning." *International Journal of Virtual and Personal Learning Environments* 4: 35–50.

National Center on Addiction and Substance Abuse at Columbia University. 2012. "The Importance of Family Dinners VIII." CASA Columbia. White Paper. Accessed on May 27, 2017. http://www.casacolumbia.org/addiction-research/reports/importance-of-family-dinners-2012.

National Council of Teachers and Mathematics (NCTM). 2014. *Principles to Actions*. Reston, VA: Author.

———. 2000. *Principles and Standards for School Mathematics*. Reston, VA: Author.

National Governors Association Center for Best Practices (NGA Center) and Council of Chief State School Officers (CCSSO). 2010. *Common Core State State Standards*. Washington, DC: NGA Center and CCSSO. http://www.corestandards.org.

National Research Council. 2001. *Adding It Up: Helping Children Learn Mathematics*. Washington, DC: National Academy Press.

Ontario Ministry of Education. 2011. "Asking Effective Questions: Provoking Student Thinking/Deepening Conceptual Understanding in the Mathematics Classroom." Capacity Building

Series, Special Edition No. 21. July http://www.edu.gov.on.ca/eng/literacynumeracy/inspire/research/CBS_AskingEffectiveQuestions.pdf.

Pagni, David, L. 2000. "An Educator's Guide to Answering Parents' Questions on Mathematics." *Teaching Children Mathematics* 7: 45–50.

Parrish, Sherry. 2010, 2014. *Number Talks: Whole Number Computation Grades K–5*. Sausalito, CA: Math Solutions.

Polya, George. 1957. *How to Solve It*. Garden City, NY: Doubleday.

Ramirez, Gerardo, Elizabeth A. Gunderson, Susan C. Levine, and Sian L. Beilock. 2013. "Math Anxiety, Working Memory and Math Achievement in Early Elementary School." *Journal of Cognition and Development* 14 (2): 187–202.

Reys, Barbara, J., and Fran Arbaugh. 2001. "Clearing Up the Confusion over Calculator Use in Grades K–5." *Teaching Children Mathematics* 8: 90–94.

Rubenstein, Rhetta N., and Denisse R. Thompson. 2002. "Understanding and Supporting Children's Mathematical Vocabulary Development." *Teaching Children Mathematics* 9: 107–12.

Schoenfeld, Alan H. 1985. *Mathematical Problem Solving*. New York: Academic Press.

Schwartz, James E. 2008. *Elementary Mathematics Pedagogical Content Knowledge: Powerful Ideas for Teachers*, Boston: Allyn & Bacon.

Seeley, Cathy. 2015. *Faster Isn't Smarter: Messages about Math, Teaching, and Learning in the 21st Century*. 2d ed. Sausalito, CA: Math Solutions.

———. 2014. *Smarter Than We Think: More Messages About Math, Teaching, and Learning in the 21st Century*. Sausalito, CA: Math Solutions.

Shetterly, Margot Lee. 2016. *Hidden Figures: The American Dream and the Untold Story of the Black Women Mathematicians Who Helped Win the Space Race*, New York: William Morrow Paperbacks.

Silver, Edward A., ed. 1985. *Teaching and Learning Mathematical Problem Solving: Multiple Research Perspectives*. Hillsdale, NJ: Lawrence Erlbaum.

Smith, Susan, S. 2013. *Early Childhood Mathematics*. New York: Pearson.

Sousa, David A. 2007. *How the Brain Learns Mathematics*. Thousand Oaks, CA: Corwin.

Stenmark, Jean Kerr. 1991. *Mathematics Assessment: Myths, Models, Good Questions, and Practical Suggestions*. Reston, VA: NCTM.

Stenmark, Jean Kerr, Virginia Thompson, and Ruth Cossey. 1986. *Family Math*. Berkeley, CA: Lawrence Hall of Science.

Stephens, Ana, Maria Blanton, Eric Knuth, Isil Isler, and Angela Murphy Gardiner. 2015. "Just Say Yes to Early Algebra!" *Teaching Children Mathematics* 22: 92–101.

Suh, Jennifer M., and Patricia S. Moyer. 2008. "Scaffolding Special Needs Students' Learning of Fraction Equivalence Using Virtual Manipulatives." *Proceedings of the International Group for the Psychology of Mathematics Education* 4: 297–304.

Sullivan, Kevin. 1992. "Foot-in-Mouth Barbie." *Washington Post*. September 30. Retrieved from www.washingtonpost.com/archive/politics/1992/09/30/foot-in-mouth-barbie/076dd0eb-996c-48f9-8050-e56ab14951b9/.

Susperreguy, Maria Ines, and Pamela E. Davis-Kean. 2016. "Maternal Math Talk in the Home and Math Skills in Preschool Children." *Early Education and Development* 27: 841–57. DOI: 10.1080/10409289.2016.1148480.

Sutton, John, and Alice Krueger. 2002. *Edthoughts: What We Know About Mathematics Teaching and Learning*. Aurora, CO: Mid-Continent Research for Education and Learning.

Tang, Greg. 2005. *Math Potatoes*. New York: Scholastic.

Van de Walle, John, Karen Karp, and Jennifer M. Bay-Williams. 2019. *Elementary and Middle School Mathematics: Teaching Developmentally*. 10th ed. Upper Saddle River, NJ: Pearson.

van Hiele, Pierre. M. 1986. *Structure and Insight: A Theory of Mathematics Education*. New York: Academic Press.

Williams, Vera B. 1982. *A Chair for My Mother*. New York: Greenwillow Books.

Yeager, David Scott, and Carol S. Dweck. 2012. "Mindsets That Promote Resilience: When Students Believe That Personal Characteristics Can Be Developed." *Educational Psychologist* 47: 302–314.

Young, Jessica Mercer, and Kristin Reed. 2017. "Mastery Motivation: Persistence and Problem Solving in Preschool" *Teaching Young Children* 11: np. Retrieved online from https://www.naeyc.org/resources/pubs/tyc/oct2017/mastery-motivation-persistence-and-problem